RANDOM PROCESSES
IN
GEOLOGY

D1694523

with contributions by

A. T. Bharucha-Reid
Department of Mathematics
Wayne State University
Detroit, Michigan, USA

I. K. Crain
Computer Science Centre
Department of Energy, Mines, and Resources
Ottawa, Canada

M. F. Dacey
Department of Geological Sciences
Northwestern University
Evanston, Illinois, USA

J. C. Davis
Kansas Geological Survey and
Department of Chemical and Petroleum Engineering
University of Kansas
Lawrence, Kansas, USA

J. E. Klovan
Department of Geology
University of Calgary
Calgary, Alberta, Canada

W. C. Krumbein
Department of Geological Sciences
Northwestern University
Evanston, Illinois, USA

F. W. Preston
Department of Chemical and Petroleum Engineering
University of Kansas
Lawrence, Kansas, USA

W. E. Price, Jr.
U. S. Geological Survey
Reston, Virginia, USA

R. A. Reyment
Paleontological Institute
Uppsala University
Uppsala, Sweden

W. Schwarzacher
Department of Geology
Queen's University
Belfast, Northern Ireland, UK

J. S. Smart
IBM Thomas J. Watson Research Center
Yorktown Heights, New York, USA

P. Switzer
Department of Statistics
Stanford University
Stanford, California, USA

F. E. Wickman
Department of Geology
University of Stockholm
Stockholm, Sweden

# RANDOM PROCESSES IN GEOLOGY

edited by

**Daniel F. Merriam**

**Springer-Verlag**
Berlin Heidelberg New York
1976

Daniel F. Merriam

Jessie Page Heroy Professor of Geology
and Chairman
Department of Geology
Syracuse University
Syracuse, New York

Library of Congress Cataloging in Publication Data

Main entry under title:

Random processes in geology.

Papers from two symposia sponsored by International Association for Mathematical Geology and held in conjunction with the International Geological Congress in Montreal, Sept. 1972.
1. Geology--Mathematics--Congresses. 2. Stochastic processes--Congresses. I. Merriam, Daniel Francis. II. International Association for Mathematical Geology. III. International Geological Congress, 24th, Montreal, 1972.
QE33.2.M3R36 550'.1'5192 75-6848



Printed in the United States of America



ISBN 0-387-07277-2 Springer-Verlag New York Heidelberg Berlin
ISBN 3-540-07277-2 Springer-Verlag Berlin Heidelberg New York

# Preface

The International Association for Mathematical Geology, in conjunction with the International Geological Congress, sponsored two symposia in Montreal, Canada, September 1972. The first symposium, Random Processes in Geology, consisted of two, half-day sessions and featured ten major papers on various aspects of stochastic models as applied to geologic problems.

The invited speakers were selected by the Projects Committee of the IAMG so as to represent a wide spectrum of geologic disciplines. The topics fell naturally into two categories: those dealing with continuous stochastic processes and those concerned with point processes and branching operations. The program, as presented, was

Introduction: R. A. Reyment
Ideal granites and their metasomatic transformation: stochastic model, statistical description, and natural rocks: A. B. Vistelius (read by G. Lea)
The influence of greisenization on the Markovian properties of grain sequences in granitic rocks: M. A. Romanova
The mechanism of bed formation in a limestone-shale environment: W. Schwarzacher
Volcanic eruptions as random events: F. E. Wickman
Statistical geometric similarity in drainage networks: J. S. Smart
Length and gradient properties of stochastic streams: M. F. Dacey
Application of stochastic point processes to volcanic eruptions: R. A. Reyment
Applications of random process models to the description of spatial distributions of qualitative geologic variables: P. Switzer
Sedimentary porous materials as a realization of a stochastic process: F. W. Preston and J. C. Davis
Stochastic process models in geology: W. C. Krumbein.

This volume consists of eight of the ten presented papers. In addition, two papers from the succeeding symposium were felt to be of sufficient interest to be included. Research in the broad field of mathematical modeling in geology is very active. It is not possible in a one-day technical meeting, or in its published proceedings, to give a full account of progress in the field. This publication, however, provides a significant collection of articles pertinent to this exciting topic.

Preface

In a complex and rapidly developing field such as that presented here, it is imperative that geologists maintain close contact with professional statisticians and mathematicians. To this end, the Projects Committee invited Professor A. T. Bharucha-Reid to attend the symposium. It is our pleasure to acknowledge gratefully his commentary and criticism, which proved to be one of the highlights of the meeting.

We wish to express our appreciation to J. E. Armstrong, C. H. Smith, S. C. Robinson, and I. M. Stevenson of the International Geological Congress Organizing Committee for their help in matters concerning the organization of the IAMG symposia.

J. E. Klovan

# Contents

# Introduction

# 1

A. T. Bharucha-Reid

An examination of the recent literature of the sciences (biological, physical, and social), engineering, and technology shows that probability theory, in particular the theory of stochastic (or random) processes, is exerting a profound influence on theoretical developments and the formulation of mathematical models in many applied fields. In turn, concrete problems posed in applied fields are motivating a considerable amount of research in probability theory.

As is well known, progress in any science is highly dependent upon developments in methodology; and there are many who believe that a highly developed science is characterized, to a great extent, by an extensive use of mathematics to formulate theories and to develop abstract models of the natural phenomena with which the particular scientific discipline is concerned. Of all the physical sciences, physics stands out as one that utilizes mathematics to analyze data and to formulate rigorous theories of physical phenomena. And, as is well known, several branches of classic and modern mathematics trace their origins to problems in physics.

Within recent years, mathematics has been playing a greater role in geology; in particular, probabilistic methods are used by an increasing number of geoscientists. Although the systematic use of probabilistic methods in geology is a relatively recent development, the geoscience literature of the early 1900s contains several papers utilizing probabilistic models. (We refer to H. O. A. Wold, 1965, Bibliography on Time Series and Stochastic Processes, MIT Press, Cambridge, Massachusetts.) In the 1940s and 1950s, studies using probabilistic methods increased, of which those of Kolmogorov (in the Soviet Union) and of Litwiniszyn (in Poland) are outstanding. The 1960s ushered in a new era with the interesting work of Dacey and Krumbein (in the USA), Schwarzacher (in the UK), Vistelius (in the Soviet Union), as well as those of other mathematical geologists.

This volume presents clear examples of concrete applications of probabilistic methods in geology. The geologist interested in mathematical geology, as well as the probabilist interested in applications of probability theory, will find a discussion of how Markov-chain processes, random-walk processes, point processes, stationary processes, and random graphs are employed in mathematical geology. As such, these papers are important contributions to a new and rapidly developing area of geology, as well as to applied probability theory.

From my examination of the literature dealing with stochastic processes in geology, I feel that the preliminary groundwork has been laid and that the probabilistic formulation of many problems in geology has been carried out correctly. At this stage, communications between mathematical geologists and probabilists should increase, and some of the more modern and powerful techniques from the theory of Markov processes, as well as other stochastic processes, should be utilized to formulate and study problems arising in geology. However, we must always be on the lookout for problems in geology that might lead to new problems in the theory of stochastic processes. That is, there must be geologic processes that require a probabilistic formulation but do not fit into (nor should they be forced into) the framework of some of the well-known classes of stochastic processes. In this manner, mathematical geology will become sophisticated, and problems in geology will lead to new developments in the theory of stochastic processes.

# Statistical Analysis of Geotectonics

# 2

I. K. Crain

ABSTRACT

Geotectonics can be approached quantitatively by applying statistical techniques to the areal distribution of tectonic elements. Linear features can be compared to random orientation and intersection distributions. The length of lineaments of stochastic origin may be described by a Gamma density. Monte Carlo simulation of several classes of random polygons allows the quantitative comparison of observed polygonal tessellations and theoretical random patterns. Simulated were the polygons caused by random lines in the plane and the Voronoi polygons. Both of these sets have application in many fields of geology.

## INTRODUCTION

The aim of this paper is to bring together some well-known, some little-known, and a few previously unknown results of geometric probability that have possible application in statistical analysis of tectonic features observable on the earth and other planets. Particular emphasis is given to tests for randomness, because the nonrandom arrangement of tectonic elements implies organization in the mode of origin, such as convection currents, or a global stress pattern. Quantitative analysis of their systematic deviations for randomness could provide information about the mechanisms involved.

It is easy to hold an intuitive concept of randomness of points in a space of any dimensions, but such intuition fails for multidimensional elements, such as lines, planes, and spheres, even in three-dimensional space. Thus, the term "random" must be defined carefully in geometric statistics, and tests for randomness refer only to the particular model considered. A rejection of an hypothesis of randomness of geometric elements therefore is a statement that the observed distribution cannot be explained readily by the specified random process. As an example, there are at least three obvious groups of random chords of a circle, each having a different mean (Kendall and Moran, 1963, p. 9-10). The physical requirements of applying geometric statistics to tectonics usually are sufficient to determine a suitable random model from among those possible.

For notation, probability density functions (density functions or densities) will be denoted by lowercase letters, the probability distribution function (distribution function, distribution, or cumulative distribution) of the same random variable by the corresponding uppercase letter. Random variables are denoted also by lowercase letters and parameters by lowercase Greek letters. For example, a normal density function with mean $\mu$ and standard deviation $\sigma$ would be denoted by $n(\mu,\sigma)$ and the corresponding cumulative distribution by $N(\mu,\sigma)$. Exceptions are made where previous convention has given general usage to a particular symbol for a density. As is common, the phrase "the probability that x is greater than or equal to g" will be contracted to $P(x\ g)$, etc. On occasion, it may be helpful to include the random variable in the notation for the density; hence, $n(x;\mu,\sigma)$. The expectation (or mean value) of a random variable x will be denoted by $E(x)$.

## RANDOMLY DISTRIBUTED POINTS

### Points in a Line

Anyone acquainted with elementary statistics is familiar with the uniform density function and the related Poisson density given by

$$\rho(n;\lambda) = \frac{\lambda^n e^{-\lambda}}{n!} \qquad n = 0, 1, 2, \ldots \tag{2.1}$$

The classic example to which such a model applies is that of a telephone switchboard where calls can be assumed to arrive uniformly randomly. There are on the average $\lambda$ calls per unit time. The probability of exactly n calls occurring in a unit time then is given by Equation 2.1. The time interval t between successive calls has the exponential density

$$f(t;\lambda) = \lambda e^{-t} \qquad t > 0 \tag{2.2}$$

If $\lambda$ is invariant with time, reference is made to a Poisson process of constant intensity; otherwise the process is of variable intensity or, in the time connotation, nonstationary (Girault, 1966, p. 8).

### Points in a Plane or Volume

For the situation of points occurring in a plane area, or in a volume, e.g., Equation (2.1), holds, $\lambda$ is the average number of points per unit area, or volume. The distance between nearest-neighbor points is no longer exponentially distributed, however. In the planar situation, the expression $2\lambda\pi r^2$ will have the $\chi^2$ distribution with two degrees of freedom (Kendall and Moran, 1963, p. 39), whereas for three dimensions, $8/3\lambda\pi r^3$ has that distribution (Kendall and Moran, 1963, p. 48). The $\chi^2$ distribution with $\gamma$ degrees of freedom is given by

$$\chi^2 \ (x;\gamma) = \frac{1}{2^{\gamma/2}\Gamma(\gamma/2)} \ \chi^{(\gamma-2)/2} \ e^{-(x/2)} \tag{2.3}$$

where $\Gamma(Z)$ denotes the tabulated function.

Many possible tests for randomness of points in a plane have been presented and applied to the distribution of crystals in a thin section (Kretz, 1969), and will not be reiterated here.

### Random Points on a Sphere

Most results and methods useful in the plane can be extended for application to a spherical surface. The obvious difference is that the area of the spherical surface is finite. Thus, some relations only hold exactly in the limiting situation of an infinite number of points, or a sphere of infinite radius. The nearest-neighbor distribution, for example, holds only if the mean interpoint distance is small compared with the radius, that is, the density of points is high. No matter how few points, however, it is true that points falling in an area of fixed size will be governed by the Poisson density (Equation 2.1).

## ANALYSIS OF LINEAR FEATURES

### Random Lines in a Plane

A line of infinite extent in the X-Y plane may be defined by two parameters $\rho$ and $\theta$ so that the equation of a general line is

$$\rho = x \cos\theta + y \sin\theta \qquad (-\infty<\rho<\infty, \ 0\leq\theta\leq\pi) \tag{2.4}$$

If $\rho$ is allowed to assume a uniformly random value between plus and minus infinity, and $\theta$ random between 0 and $\pi$, a set of random lines is produced ($\rho$ can be thought of as a distance from an arbitrary origin). Figure 2.1 shows an example of part of such a set.

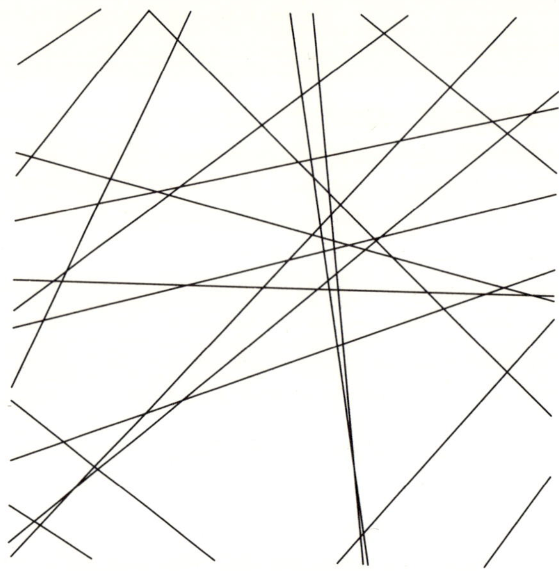

FIGURE 2.1 Typical representation of random lines in plane.

Because there is an obvious correspondence between these random lines and random points in an area, or rather in infinite strip of width $\pi$, many results of the previous section apply: the distance from an arbitrary origin to a line will be uniformly randomly distributed between 0 and infinity; the azimuths of the lines also are uniformly distributed. Other results are given by Miles (1964a) as follows. The mutual intersection angles of the lines (taken as the least of the two complementary angles) have the probability density function $\sin\theta$ $(0\leq\theta\leq\pi/2)$. This is a fairly important result, as, at first glance, one might expect a uniform distribution, whereas in fact as intersection is more probable near $90^o$ than near $0^o$.

The number of lines intersecting any convex figure of perimeter s has a Poisson density with mean $\tau s/\pi$ where $\tau$ is the mean number of lines per unit length. This is an additional useful test for randomness, similar to the Poisson density test used for points in an area.

Random Great Circles on a Sphere

Most results of the previous section apply directly to the situation of great circles on a sphere. The particular great circle is characterized by two angles between 0 and $\pi$ that assume uniformly random values independently of each other. Two exactly equivalent intersections of two great circles will occur at conjugate points on the sphere. For this reason, measurements need only be made on one hemisphere. Otherwise the distribution of intersection angles and number of lines intersecting convex figures is the same.

Length of Linear Features

Analysis of lineaments as an aid to interpreting regional tectonics has been widely used. Most definitions of the word "lineament" (see, for example, Dictionary of Geological Terms, 1960; or Dennis, 1967) require a topographic alignment known to be structurally controlled. It is obvious, however, that a linear structural feature need not have a consistent topographic expression, and further that linear arrangements of entirely unrelated features may occur owing to the chance alignment of more or less random geologic events. In this work, the term "lineament" will be used broadly for any approximate rectilinear alignment of geologic, tectonic, or topographic features of known or unknown origin, whether corelated or not.

As opposed to an infinite line, a lineament, being a line segment, has a third measurable parameter, its length. Because the length distribution of a set of lineaments may reflect gross tectonics, it is important to consider possible random distributions, or, more correctly, stochastic models for the distribution of lineament lengths.

In determining the appropriate stochastic model one must consider the various factors that influence the decision to draw a lineament on a geologic or topographic map, for example, to interrupt or continue the line. The assemblage of lineaments a person determines from a map is the result of interaction among his processes of visual perception, the scale, resolution and accuracy of the map, and the actual physical location of the points and distinct areas on the map. Because it is impossible to separate these psychological and geometric effects, formulation of a mathematical model paralleling this process is difficult. Some consideration of the perception aspects of this problem is given by Julesz (1962).

Intuitively, both long and short lineaments will be uncommon; there will be none of negative length and the mean will exceed the (single) mode. One family of such positively skewed densities is the Gamma density, also called the Person Type III, (Kendall and Stuart, 1961, p. 64), given by

$$g\ (x;\alpha,\beta) = \frac{x^{\alpha}}{\Gamma(\alpha)\beta^{\alpha}}\ e^{-x/\beta} \tag{2.5}$$

Obviously, testing of observed lineaments for randomness against a Gamma density is conservative. It is possible for lineaments of nonstochastic origin, owing to a single physical phenomenon, to show a Gamma density, but, on the other hand, truly random lineaments caused by disorganized sources will be highly unlikely to deviate significantly from a Gamma density. Thus the test is sufficient but not necessary: Whereas a significant deviation from a Gamma density is evidence of nonrandomness, the converse is not necessarily true.

## DISTRIBUTION OF POLYGONAL ELEMENTS

### Polygons Formed by Random Lines in a Plane

The set of random lines described in the previous section delineates an assemblage of polygons. Of interest are the density functions of the number of sides $n$, the lengths of the sides $l$, the perimeter $s$, the in-circle (inscribed circle of largest diameter) $d$, and the area $a$ of the polygons. The major references to what has been done in this field are Goudsmit (1945) and Miles (1964a, 1964b).

The following results are known (Miles, 1964a, p. 904):

The density of $d$ is exponential with mean $1/\tau$
The density of $2\tau s/\pi$ is exponential with mean 1.

The expectation values of some of the variables are as follows:

$$E(n) = 4$$
$$E(s) = 2\pi/\tau$$
$$E(a) = \pi/\tau^2$$
$$E(n^2) = (\pi^2 + 24)/2$$
$$E(s^2) = \pi^2(\pi^2 + 4)/2\tau^2$$
$$E(a^2) = \pi^2/2\tau^4$$
$$E(a^3) = 4\pi^7/7\tau^6$$

These variables, of course, are not independently distributed, as, for example, the number of sides of large polygons probably will be higher than on small polygons. Thus the cross expectations also are important (Miles, 1964a, p. 904):

$$E(sn) = \pi(\pi^2 + 8)\ /2\tau,\ E(an) = \pi^3/2\tau^2,\ E(as) = \pi^4/2\tau^3$$
$$E(a^2n) = \pi^4(8\pi^2 - 21)/21\tau^4,\ E(a^2s) = 8\pi^7/21\tau^5$$
$$E(a^{m-1}s) = \frac{2\tau^{E(a^m)}}{m}$$

Although the exact density of none of the variables is known, the probability that $n = 3$ has been calculated to be 0.3551 (Miles, 1964a, p. 903) and the density of $2\tau s/\pi$ for the class of k-sided polygons is $\chi^2$ with $2(k - 2)$ degrees of freedom. Thus, if the expected frequencies of the number of sides were known to be $p_3$, $p_4$, $p_5$, . . ., then the density of $2\tau s/\pi$ would be

$$f(s) = \sum_{i=1}^{\infty} p_k \chi^2_{2(k-2)} \tag{2.6}$$

Although certain tests may be applied with only the knowledge of the means and other expectations, obviously it is of great interest to obtain the densities more explicitly. The author has obtained values of $p_k$ by a Monte Carlo method that is detailed elsewhere (Crain and Miles, 1975).

These values are shown in Table 2.1 with the indicated possible error (95% confidence). Figure 2.2A shows the density and distribution function of $2\tau s/\pi$. Figure 2.2B shows a histogram of the frequencies of areas of polygons obtained by the same method.

Polygons Formed by Random Great Circles on a Sphere

The fundamental difference between random great circles and random lines is that in the latter the area under consideration is infinite, whereas in the former it is finite. For instance, it is ridiculous to consider the distribution of areas of the figures formed by two random lines in a plane, because all these figures will be infinite in extent. On the other hand, two great circles form four (two pairs) infinite figures. In fact, it is absurd to consider any finite number of lines in a plane. Rather, one discusses an infinite number with average intensity $\tau$. On the sphere, however, all expectation values differ with the number (finite) of great circles. There is an obvious correspondence, however. If on a unit sphere there are randomly placed N great circles, one would encounter on a random traverse of the globe 2N intersections. Although it is difficult to prove rigorously, all results of the previous section can

TABLE 2.1 $p_n$ for Poisson polygons

| n | Frequency | $p_n$ |
|---|---|---|
| 3 | 36,732 | 0.3551[a] |
| 4 | 37,857 | 0.3779 |
| 5 | 18,795 | 0.1918 |
| 6 | 5,672 | 0.0592 |
| 7 | 1,233 | 0.0132 |
| 8 | 179 | 0.00196 |
| 9 | 24 | 0.00025 |
| 10 | 3 | 0.000038 |

[a]Known values of all others estimated by Monte Carlo methods.

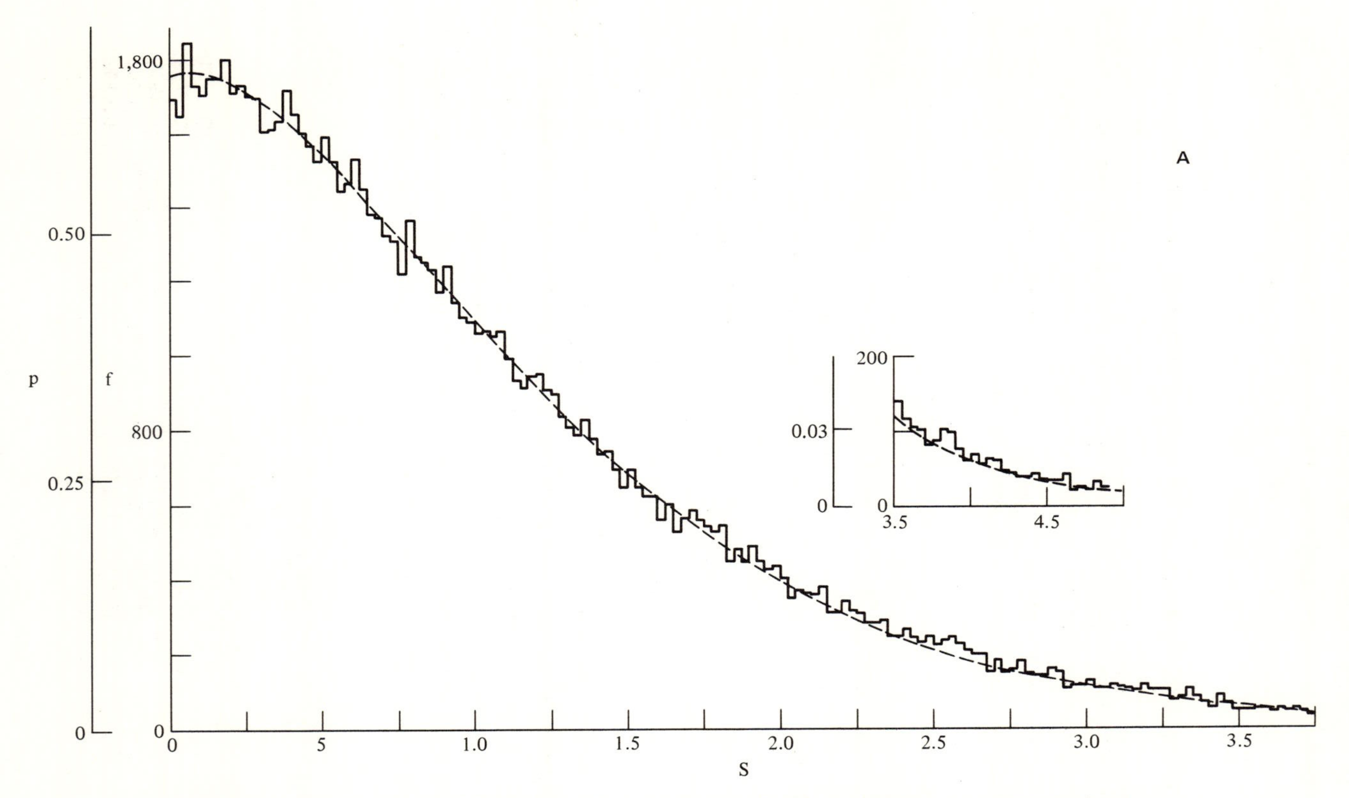


FIGURE 2.2A Frequency histogram of perimeters of Poisson polygons. Inset shows right-tail of distribution. Approximate probability density scale (p) and frequency scale (f) are shown.

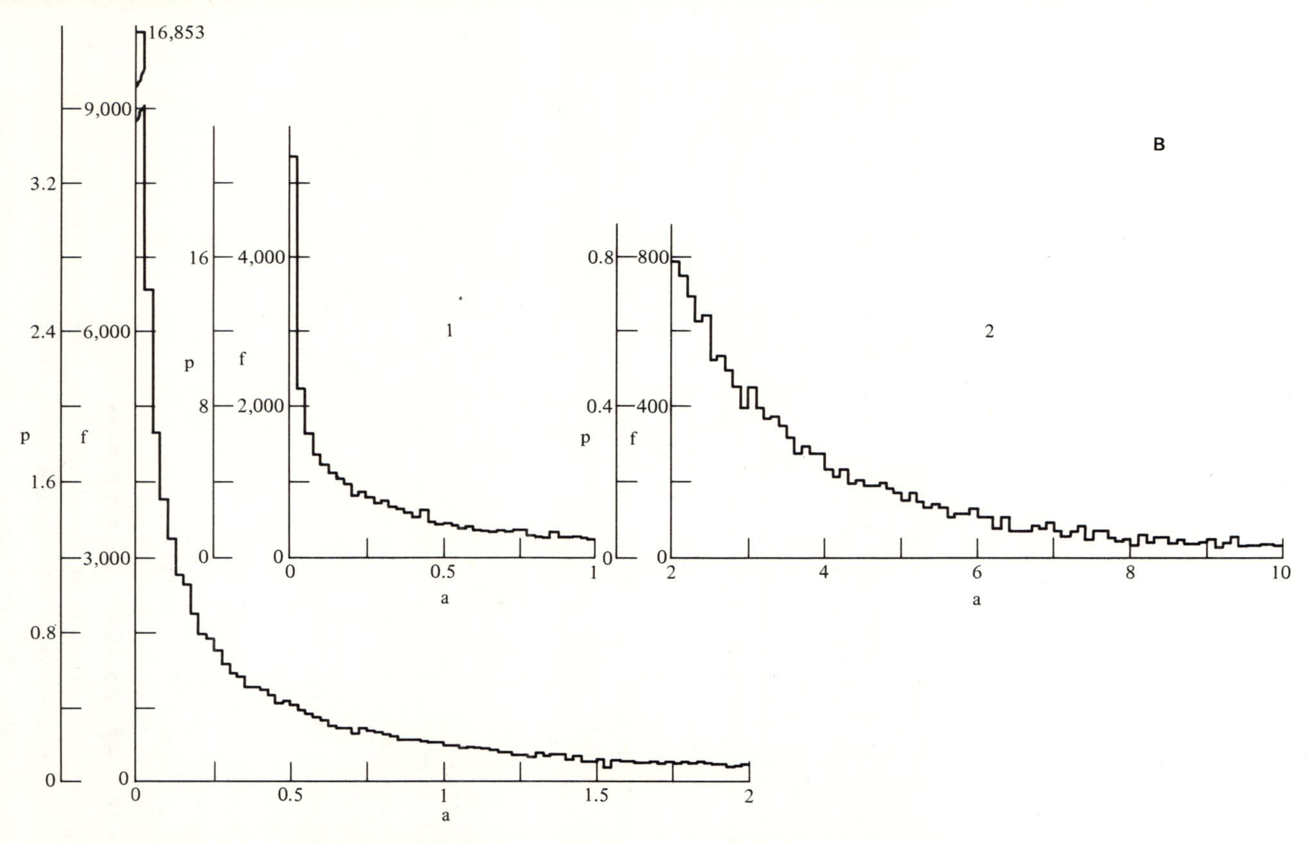


FIGURE 2.2B Frequency histogram of areas of Poisson polygons. Inset 1 shows detail of left-tail of distribution; Inset 2 shows detail of right-tail. Approximate probability density scale (p) and frequency scale (f) are shown.

be applied to the sphere, for sufficiently large N, by replacing $\tau$ with 2N. For small N, this is not true; for example, the mean number of sides of the spherical polygons is not 4, but rather (Miles, 1971)

$$E_N(n) = \frac{4N(N-1)}{N^2 - N + 2} \tag{2.7}$$

Similarly

$$E_N(s) = \frac{4\pi N}{N^2 - N + 2}$$

and

$$E_N(a) = \frac{4\pi}{N^2 - N + 2}$$

These expressions soon approach the planar values for relatively low N, for example, $E_{100}(n) = 3.992$.

One important additional property of these distributions is that they are invariant if through a random (or nonrandom) process some of the lines, or parts of lines, are covered, removed, or thickened (Miles, 1969a). Those polygons observed to be complete will have the same statistical properties as the entire set. This is of obvious importance in tectonic analyses where the preserved record of tectonic lineaments is fragmentary. It is equally valid, in the statistical sense, to extend great circle segments to full closure and to perform tests on this hypothetical set of polygons.

### The Voronoi Polygons

The class of random polygons that describes problems of growth about random centers, or the contraction cracking of a surface, is the Voronoi polygon (Gilbert, 1962). Figure 2.3 shows a typical type of tessellation; known results are (Miles, 1970)

$$E(n) = 6$$
$$E(s) = 4/\rho^{\frac{1}{2}}$$
$$E(a) = 1/\rho$$
$$E(a^2) = 1.280/\rho^2$$

where $\rho$ is the intensity of the Poisson process.

Area, perimeter, and side distributions were previously unknown. The author has performed a Monte Carlo simulation to estimate these distributions. Table 2.2 shows the approximate probabilities of the various-sided polygons; Figures 2.4A and 2.4B show the area and perimeter distributions.

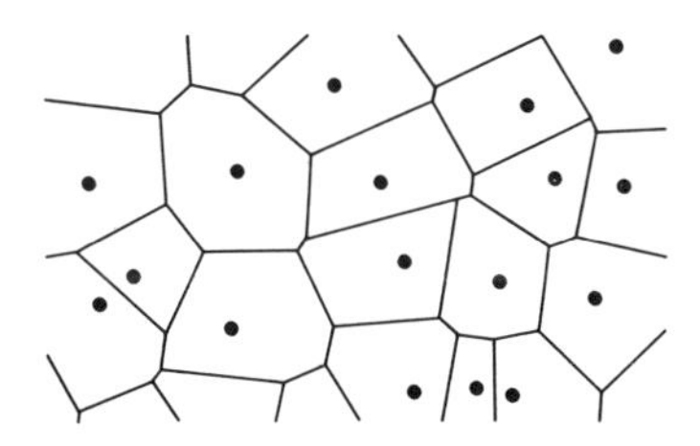

FIGURE 2.3 Typical representation of Voronoi polygons.

TABLE 2.2 Probabilities of sides of Voronoi polygons

| n | Frequency | $p_n$ |
|---|---|---|
| 3 | 125 | 0.011 |
| 4 | 1,215 | 0.110 |
| 5 | 2,846 | 0.259 |
| 6 | 3,172 | 0.288 |
| 7 | 2,266 | 0.206 |
| 8 | 953 | 0.087 |
| 9 | 321 | 0.029 |
| 10 | 85 | 0.0077 |
| 11 | 15 | 0.0014 |
| 12 | 2 | 0.0002 |

These distributions allow for the testing of expansion-crack hypotheses for tectonics on a regional or global scale.

## CONCLUDING REMARKS

It is impossible in a short paper to give practical examples of the application of these techniques. The author has applied them successfully to terrestrial and planetary tectonics. Possible areas of application include the analyses of length, orientation, and intersection angle of large-scale linear features, and the quantitative evaluation of various hypothetical global polygon networks that various workers have proposed.

## REFERENCES

Anonymous, 1960, Dictionary of geological terms (2nd ed.): Am. Geol. Inst., Doubleday & Co., Garden City, New York, 545 p.

Crain, I. K., and Miles, R. E., 1975, Monte Carlo estimates of frequency distributions due to random lines in the plane: in preparation.

Dennis, J. G., ed., 1967, International tectonic dictionary: Am. Assoc. Petroleum Geologists, Tulsa, Oklahoma, 196 p.

Gilbert, E. N., 1962, Random subdivisions of space into crystals: Ann. Math. Stat., v. 33, no. 3, p. 958-972.

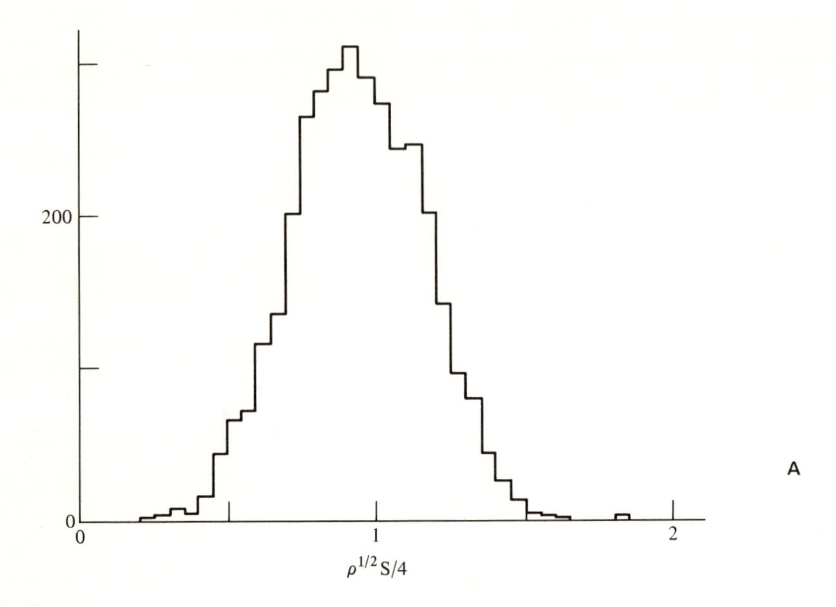


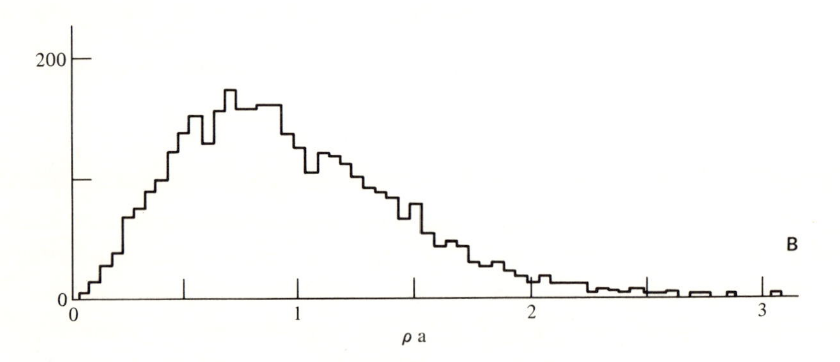


FIGURE 2.4 A. Frequency histogram of perimeters of random Voronoi polygons (5,000 polygons). B. Frequency histograms of areas of same polygons.

Girault, M., 1966, Stochastic processes: Springer-Verlag, Berlin, 125 p.

Goudsmit, S., 1945, Random distribution of lines in a plane: Rev. Mod. Phys., v. 17, no. 2, p. 321-322.

Julesz, B., 1962, Visual pattern recognition: IRE Trans. Prof. Group Inform. Theory, v. IT-8, no. 2, p. 1-9.

Kendall, M. G., and Moran, P. A. P., 1963, Geometrical probability: Charles Griffen and Co., London, 125 p.

Kendall, M. G., and Stuart, H., 1961, The advanced theory of statistics, V. 2: Charles Griffen and Co., London, 690 p.

Kretz, R., 1969, On the spatial distribution of crystals in rocks: Lithos, v. 2, no. 1, p. 39-66.

Miles, R. E., 1964a, Random polygons determined by random lines in a plane. I: Proc. Natl. Acad. Sci. (USA), v. 52, no. 4, p. 901-907.

Miles, R. E., 1964b, Random polygons determined by random lines in a plane. II: Proc. Natl. Acad. Sci. (USA), v. 52, no. 5, p. 1157-1160.

Miles, R. E., 1970, On the homogeneous planar Poisson point process: Math. Biosci., v. 6, no. 1, p. 85-127.

Miles, R. E., 1971, Random points, sets, and tessellations on the surface of a sphere: Sankhya, Ser. A, v. 33, no. 2, p. 145-174.

# 3

# Summary of Magnitude Properties of Topologically Distinct Channel Networks and Network Patterns

M. F. Dacey

ABSTRACT

This paper reviews and summarizes properties of the magnitude of links that comprise topologically random channel networks and network patterns. Particular emphasis is given to random variables that are defined by (1) the magnitude of a link randomly selected from the links comprising all topologically distinct channel networks of magnitude n, (2) the magnitude of a link randomly selected from the links comprising all topologically distinct network patterns of magnitude n, and (3) the number of channel networks in a network pattern randomly selected from all topologically distinct network patterns of magnitude n. Consideration also is given to a classification of links that depends upon both the magnitude of a link and its position in a channel network.

## INTRODUCTION

Although there is much empirical and theoretical literature on quantitative aspects of networks and basins that form drainage systems, the major conceptual contributions are attributable to a few basic studies. The empirical quantitative study of drainage systems was initiated mainly by Horton's (1945) specification of the concept of stream order and statement of two fundamental laws of drainage composition that connect the numbers and lengths of the streams of different orders in a drainage network. Strahler (1952) removed the ambiguity from the definition of stream order and opened the investigation of drainage systems to mathematical analysis, and Shreve's (1966) concept of link magnitude provides a structure that seemingly is more amenable to formal analysis. In addition, the relation to stochastic processes was provided by Shreve's (1966, 1967) identification of the concept of topologically random channel networks, which provides a firm probabilistic basis for the statistical analysis of drainage systems. Werner's (1971) identification of topologically random network patterns provides a more general stochastic formulation in which the topologically random channel network is a special situation.

The nonprobabilistic analysis of drainage systems that is based upon the Strahler definition of order is reviewed adequately and summarized by Strahler (1964) and Morisawa (1968). This review paper concentrates on the magnitude properties that derive from the formulation by Shreve and Werner of topologically random network structures. Particular emphasis is given to the identification of random variables that describe the magnitude properties of channel networks and network patterns. This identification of properties includes a modification of Mock's (1971) classification of a link on the basis of its magnitude and position in a channel network.

The first section provides basic definitions and the structure used for the study of channel networks and network patterns. The next section identifies properties of random variables currently used in the study of this structure. The last section considers the classification of links. Although some new results are given, a primary purpose of this paper is simply to list in one place some of the basic properties of link magnitude. Proofs are not given for these properties, but in appropriate places I have tried to identify where the result first occurs in the study of channel networks. Proofs for these and many additional properties are given elsewhere (Dacey, 1972c).

## BASIC PROPERTIES OF CHANNEL NETWORKS AND NETWORK PATTERNS

This study uses the concepts, terminology, and results of graph theory, although some of the terminology is modified so as to be consistent with terms commonly used in the study of channel networks. In the terminology of graph theory, a channel network is a special type of graph consisting of a collection of edges and vertices that form a planted plane tree in which each vertex has valency 1 or 3. This graph is formulated in this study as a collection of "links" and "nodes."

DEFINITION 3.1. The two nodes of each link are distinguished as "upnode" and "downnode." A "fork" is formed by the coincidences of nodes of three distinct links - the downnodes of two distinct links and the upnode of a third link - and these three nodes are called "members" of a fork. The two links whose downnodes are members of a fork are called the "branches" of a fork, and these branches are oriented with respect to the third link and are distinguished as the "left branch" and the "right branch" of the fork. A "nodal point" is an isolated node that does not coincide with any other link. An upnode (downnode) that is a nodal point is called a "source" (outlet). A "path" is a sequence of one or more links in which no link occurs more than once. Figure 3.1 illustrates some of these concepts.

DEFINITION 3.2. A channel network of magnitude $n \geq 1$ is a collection $\lambda_n$ of links, along with the resulting forks and nodal points, that have the following properties.

1. Each node of every link in $\lambda_n$ is a source, an outlet, or a member of one fork.
2. There are exactly n links in $\lambda_n$ that have an upnode that is a source and exactly one link in $\lambda_n$ that has a downnode that is an outlet.
3. There is exactly one path between every pair of nodes of links in $\lambda_n$.

DEFINITION 3.3. The links in any collection $\lambda_n$ are classified in the following manner. An "exterior link" is a link whose upnode is a source. An "interior link" is a link whose upnode is a member of a fork. A "terminal (or outlet) link" is a link whose downnode is an outlet.

DEFINITION 3.4. Each link in a channel network is assigned a "magnitude." An exterior link is said to have a magnitude 1. The magnitude of an interior link is equal to the sum of magnitudes of the two links that form a fork at its upnode. Figure 3.2 illustrates this concept.

DEFINITION 3.5. Two-channel networks of magnitude n are called "isomorphic" if there is a one-to-one mapping of the nodes of one

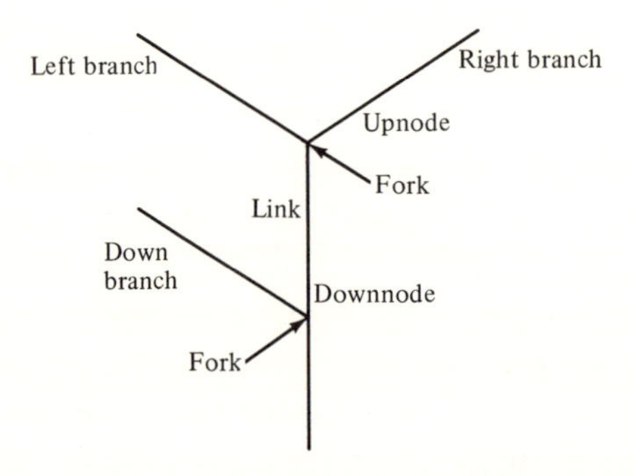


FIGURE 3.1 Example of terms defined with respect to link.

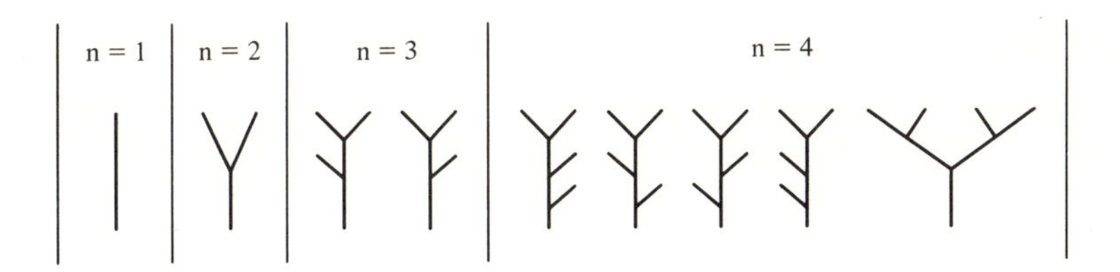


FIGURE 3.2 Examples of topologically distinguishable channel networks comprising collection $\Lambda_n$ for $1 \leq n \leq 4$.

onto those of the other that preserves adjacency. Two-channel networks of magnitude n are called "map-isomorphic" if there is an isomorphism that preserves the terminal link and the cyclic order of links at corresponding nodes. Two-channel networks are called "distinguishable" if they are of different magnitudes or if they are of the same magnitude and are not map-isomorphic. Distinguishable channel networks also are called "topologically distinct." Figure 3.3 illustrates the topologically distinct channel networks of small magnitude.

DEFINITION 3.6. The following notation is utilized:

$\lambda_n$ is a channel network of magnitude n.

$\Lambda_n$ is the collection of topologically distinct channel networks of magnitude n.

N(n) is the number of channel networks in $\Lambda_n$.

M(n) is the number of links in $\Lambda_n$.

M(k,n) is the number of magnitude k links in $\Lambda_n$.

THEOREM 3.1. A channel network $\lambda_n$ consists of 2n - 1 links. The functions N(n), M(n) and M(k,n) that describe the collection $\Lambda_n$ have the values given in Table 3.1.

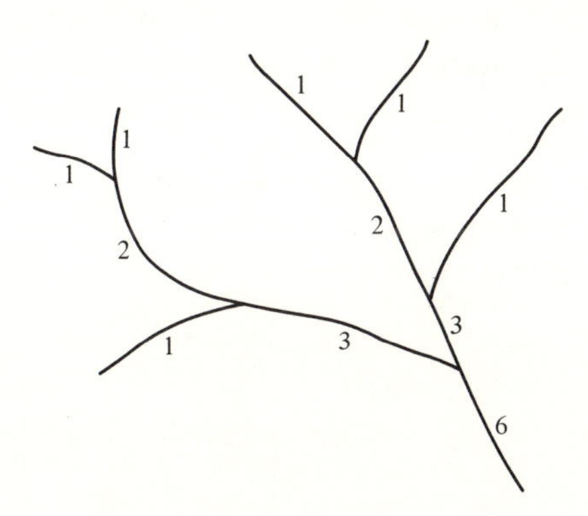


FIGURE 3.3 Example of assignment of link magnitude; (link) length of this channel network is 4.

TABLE 3.1 Properties[a]

| | Function | Source |
|---|---|---|
| (1) | $N(n) = \frac{(2n-2)!}{(n-1)!n!}$ | Shreve (1966) |
| (2) | $M(n) = \binom{2n-1}{n}$ | Shreve (1967) |
| (3) | $M(k,n) = \frac{(2k-2)!}{(k-1)!k!} \binom{2n-2k}{n-k}$ | Shreve (1967) |
| (4) | $R(n) = \binom{2n+1}{n-1}$ | Dacey (1972a) |
| (5) | $R(k,n) = \frac{(2k-2)!}{(k-1)!k!} \binom{2n-2k+2}{n-k}$ | Dacey (1972a) |
| (6) | $S(n) = \frac{3(2n)!}{(n-1)!(n+2)!}$ | Werner (1971) |
| (7) | $S(k,n) = \frac{(2n-2k+2)!}{(n-k+1)!(n-k+2)!} \frac{(2k-2)!}{(k-1)!k!}$ | Werner (1971) |
| (8) | $T(n) = \frac{(2n)!}{n!(n+1)!}$ | Werner (1971) |
| (9) | $T(k,-) = \frac{k(2n-k-1)!}{(n-k)!n!}$ | Dacey (1971) |

[a]Unless otherwise noted, $n$ is a positive integer and $1 \leq k \leq n$.

The concept of network patterns was introduced into the study of channel networks by Werner (1971). The following formulation of network pattern is adapted from that of Dacey (1971).

DEFINITION 3.7. A network pattern of magnitude n is an ordered collection of one or more disjoint channel networks such that n is the sum of magnitudes of its component channel networks.

DEFINITION 3.8. Let $\lambda_{n(i)}$ represent a channel network of magnitude $n_i$. A network pattern $\omega_{k,n}$ of magnitude n that contains k channel networks is completely enumerated by the collection

$$\{\lambda_{n(1)}; \lambda_{n(2)}, \ldots, \lambda_{n(k)} : n_1 + n_2 + \ldots + n_k = n\}$$

$$= \left\{ \bigcup_{i=1}^{k} \lambda_{n(i)} : n \right\} = \omega_{k,n}$$

where the order of the $\lambda_{n(i)}$ indicates the left to right ordering of the channel networks comprising $\omega_{k,n}$.

DEFINITION 3.9. Consider the network pattern $\omega_{k,n}$ and another network pattern

$$\omega^!_{j,n} = \left\{ \bigcup_{i=1}^{j} \lambda'_{n(i)} : n \right\}$$

The network patterns $\omega_{k,n}$ and $\omega^!_{j,n}$ are called topologically distinct network patterns of magnitude n wherever $k \neq j$ or $k = j$ and at least one of the channel networks $\lambda_{n(i)}$ is topologically distinct from $\lambda'_{n(i)}$. Figure 3.4 illustrates this concept.

DEFINITION 3.10. The following notation is utilized.

$\Omega_n$ is the collection of topologically distinct network patterns of magnitude n.

R(n) is the number of links in $\Omega_n$.

R(k,n) is the number of magnitude k links in $\Omega_n$.

S(n) is the number of channel networks in $\Omega_n$.

S(k,n) is the number of magnitude k channel networks in $\Omega_n$.

T(n) is the number of network patterns in $\Omega_n$.

T(k,n) is the number of network patterns in $\Omega_n$ that consist of k channel networks.

THEOREM 3.2. The six preceding functions that describe the collection $\Omega_n$ have the values given in Table 3.1.

## RANDOM VARIABLES FOR CHANNEL NETWORKS AND NETWORK PATTERNS

This section identifies random variables that describe properties of the collections of topologically distinct channel networks and network patterns. Channel networks of magnitude n are described in terms of the random variable $X_n$ that represents the magnitude of a link randomly selected from the links comprising the channel network in $\Lambda_n$. Network patterns of magnitude n are described in terms of the random variable $Y_n$ that represents the magnitude of a link randomly selected from the links comprising the network pattern in $\Omega_n$ and the random variable $Z_n$ that represents the number of channel networks comprising a network pattern that is selected randomly from the network pattern in $\Omega_n$. Also considered are the limiting forms of these three random variables obtained by letting n tend to infinite. The major results are summarized in Tables 3.2 and 3.3.

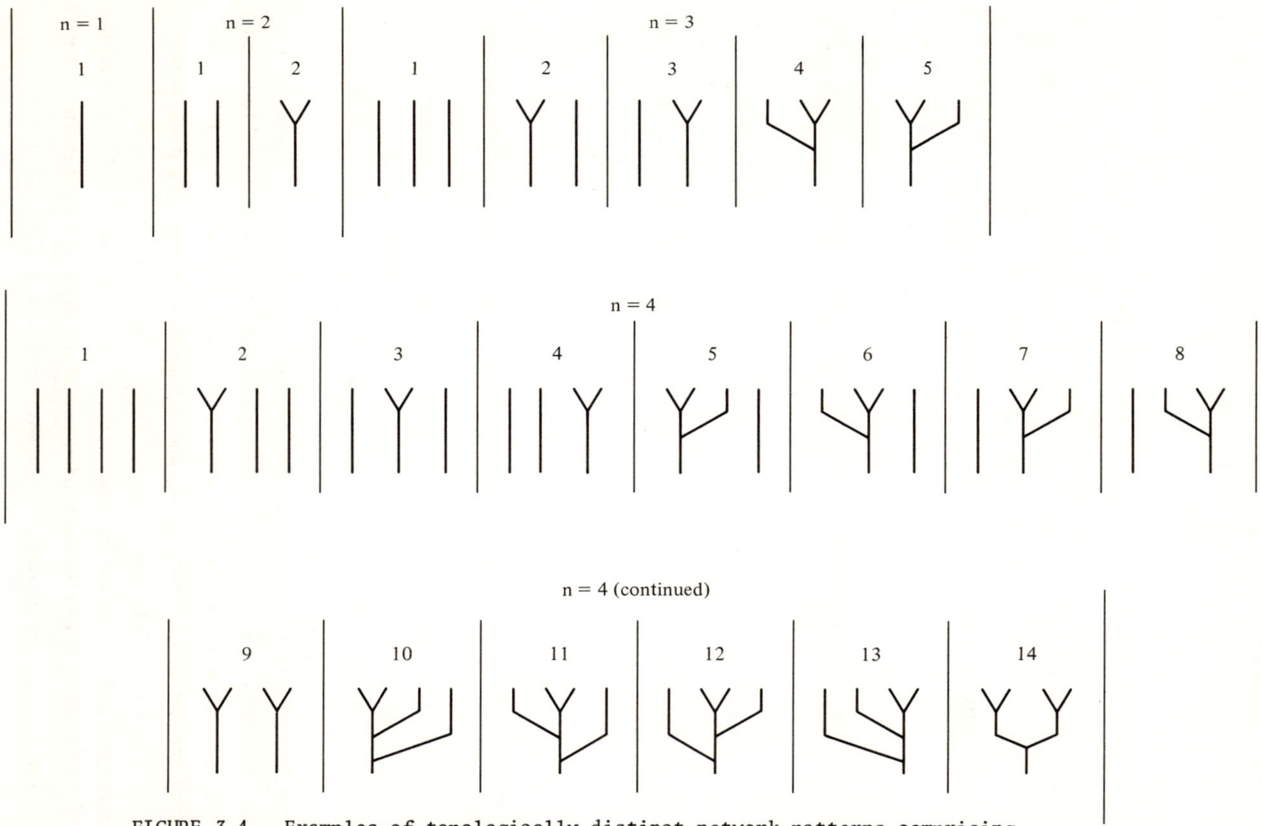

n = 1

1

n = 2

1 2

n = 3

1 2 3 4 5

n = 4

1 2 3 4 5 6 7 8

n = 4 (continued)

9 10 11 12 13 14

FIGURE 3.4 Examples of topologically distinct network patterns comprising collection $\Omega_n$ for $1 \leq n \leq 4$.

DEFINITION 3.11. The generalized hypergeometric series function of $p + q$ parameters is

$$ {}_pF_q[a_1, \ldots, a_p; c_1, \ldots, c_q; z] = \sum_{i=0}^{\infty} \frac{\Gamma(i + a_1) \ldots \Gamma(i + a_p)}{\Gamma(a_1) \ldots \Gamma(a_p)} \frac{\Gamma(c_1) \ldots \Gamma(c_q)}{\Gamma(i + c_1) \ldots \Gamma(i + c_q)} \frac{z^i}{i!} $$

where $a_i$ and $c_i$ are not negative integers. The terminating generalized hypergeometric series function of $n + 1$ terms and $p + q + 1$ parameters is

$$ {}_{p+1}F_q[-n, a_1, \ldots, a_p; c_1, \ldots, c_q; z] $$

$$ = \sum_{i=0}^{\infty} \frac{\Gamma(i - n)}{\Gamma(-n)} \frac{\Gamma(i + a_1) \ldots \Gamma(i + a_p)}{\Gamma(a_1) \ldots \Gamma(a_p)} \frac{\Gamma(c_1) \ldots \Gamma(c_q)}{\Gamma(i + c_1) \ldots \Gamma(i + c_q)} \frac{z^i}{i!} $$

where $a_i$ and $c_i$ are not negative integers or any $a_i$ or $c_i$ that is a negative integer has value less than $-n$.

Properties of the generalized hypergeometric series function are given, for example, by Erdelyi (1953).

A large and important family of random variables has generating functions that may be expressed as the quotient of two generalized hypergeometric series functions, and these random variables occur frequently in the analysis of channel networks and network patterns. Properties of this family of random variables are given by Kemp (1968) and Dacey (1972a).

THEOREM 3.3. Let $X_n$ represent the magnitude of a link randomly selected from the links comprising the channel networks in $\Lambda_n$. Then the generating function $E\ t^{X_n}$, the probability distribution $P\{X_n = k\}$, and the rth factorial moment $E\ X_n^{(r)}$ are as given in Table 3.2.

THEOREM 3.4. Put $X = \lim_{n \to \infty} X_n$. The generating function $E\ t^X$ and the probability distribution $P\{X = k\}$ are as given in Table 3.3. Moreover, the expected value $E\ X^r = \infty$ for all positive integer $r$.

THEOREM 3.5. Let $Y_n$ represent the magnitude of a link randomly selected from the links in $\Omega_n$. Then the generating function $E\ t^{Y_n}$, the probability distribution $P\{Y_n = k\}$, and the rth factorial moment $E\ Y_n^{(r)}$ are as given in Table 3.2.

THEOREM 3.6. Put $Y = \lim_{n \to \infty} Y_n$. Then $Y = X$.

TABLE 3.2 Random variables[a]

| | Function | | Source |
|---|---|---|---|
| | Definitions and Probability Distributions | | |
| (1) | $P\{X_n = k\} = M(k,n)/M(n) = \frac{(2k-2)!}{(k-1)!k!}\binom{2n-2k}{n-k}\binom{2n-1}{n}^{-1}$ | | Shreve (1966) |
| (2) | $P\{Y_n = k\} = R(k,n)/R(n) = \frac{(2k-2)!}{(k-1)!k!}\binom{2n-2k+2}{n-k}\binom{2n+1}{n-1}^{-1}$ | | Dacey (1972a) |
| (3) | $P\{Z_n = k\} = T(k,n)/T(n) = \frac{k(2n-k-1)!}{(n-k)!}\frac{(n+1)!}{(2n)!}$ | | Dacey (1971) |
| | Generating Functions | | |
| (4) | $E\,t^{X_n} = ct\,{}_3F_2[-n+1, \frac{1}{2}, 1; -n+3/2, 2; t],\ c = \frac{n}{2n-1}$ | | Dacey (1972a) |
| (5) | $E\,t^{Y_n} = ct\,{}_4F_3[-n+I, -n-1, \frac{1}{2}, 1; -n, -n+\frac{1}{2}, 2; t],\ c = \frac{n+2}{2n+1}$ | | Dacey (1972a) |
| (6) | $E\,t^{Z_n} = ct\,{}_2F_1[-n+1, 2; -2n+2; t],\ c = \frac{n+1}{2(2n-1)}$ | | Dacey (1971) |
| | Factorial Moments | | |
| (7) | $E\,X_n^{(r)} = 2^{2n-2}\binom{2n-1}{n}^{-1}$ | $r = 1$ | |
| | $= 2^{2n-2r+1}\frac{(2r-3)!}{(r-2)!}\binom{n-1}{r-1}\binom{2n-1}{n}^{-1}$ | $2 \leq r \leq n$ | |
| (8) | $E\,Y_n^{(r)} = 2\frac{(2n-2r+2)!(2r-3)!}{(n-r)!(n-r+2)\ (r-2)!}\binom{2n+1}{n-1}^{-1}$ $\cdot\ {}_3F_2[-n+r, -n+r-2, r-\frac{1}{2}; -n+r-1, -n+r-\frac{1}{2}; 1]$ | $1 \leq r \leq n$ | |
| (9) | $E\,Z_n^{(r)} = (2r+1)\frac{n!(n+1)!r!}{(n-r)!(n+r+1)!}$ | $1 \leq r \leq n$ | |

TABLE 3.2 (Continued)

| | Special Values |
|---|---|
| (10) | $E\,X_n = 2^{2n-2} \binom{2n-n}{n}^{-1}$ $\quad P\{X_n = 1\} = \frac{n}{2n-1} \geq \frac{1}{2}$ |
| | $V\,X_n = \frac{2^{2n-4}}{(2n-1)!} \binom{2n-1}{n}^{-1} [2(n+1)(2n-1)! - 2^{2n} n!(n-1)!]$ |
| (11) | $E\,Y_n = 2^{2n} \binom{2n+1}{n-1}^{-1} - \frac{n+2}{n}$ $\quad P\{Y_n = 1\} = \frac{n+2}{2n+1} \geq \frac{1}{2}$ |
| (12) | $E\,Z_n = \frac{3n}{n+2} \leq 3$ $\quad P\{Z_n = 1\} = P\{Z_n = 2\} \geq \frac{1}{4}$ |
| | $V\,Z_n = \frac{2n(n-1)(2n+1)}{(n+2)(n+2)(n+3)} \leq 4$ |

[a]Unless otherwise stated, r and n are positive integers and $1 \leq k \leq n$.

TABLE 3.3 Limiting form of random variables[a]

| | Function | Source |
|---|---|---|
| | Definitions and Probability Distributions | |
| (1) | $P\{X = k\} = \lim_{n \to \infty} P\{X_n = k\} = 2^{-(2k-1)} \frac{(2k-2)!}{(k-1)!k!}$ | Shreve (1967) |
| (2) | $P\{Y = k\} = \lim_{n \to \infty} P\{Y_n = k\} = 2^{-(2k-1)} \frac{(2k-2)!}{(k-1)!k!}$ | |
| | $= P\{X = k\}$ | Dacey (1972a) |
| (3) | $P\{Z = k\} = \lim_{n \to \infty} P\{Z_n = k\} = k(\frac{1}{2})^{k+1}$ | Dacey (1971) |
| | Generating Functions | |
| (4) | $E\ t^X = ct\,{}_2F_1[\frac{1}{2}, 1; 2; t]$   $c = \frac{1}{2}$ | Dacey (1972a) |
| (5) | $E\ t^Y = ct\,{}_2F_1[\frac{1}{2}, 1; 2; t]$   $c = \frac{1}{2}$ | Dacey (1972a) |
| (6) | $E\ t^Z = t\,{}_1F_0[2; \frac{1}{2}t]/{}_1F_0[2; \frac{1}{2}] = t(2-t)^{-2}$ | Dacey (1971) |
| | Factorial Moments | |
| (7) | $E\ X^{(r)} = +\infty$ | |
| (8) | $E\ Y^{(r)} = +\infty$ | |
| (9) | $E\ Z^{(r)} = (2r+1)r!$ | |
| | Special Properties | |
| (10) | $P\{X = 1\} = \frac{1}{2}$ | |
| | $P\{X = k+1 \mid X > k\} = \frac{1}{2(k+1)}$ | |
| (11) | $E\ Z = 3$   $V\ Z = 4$   $P\{Z = 1\} = P\{Z = 2\} = \frac{1}{4}$ | |
| | $P\{Z = k+1 \mid Z > k\} = \frac{k+1}{2(k+2)}$ | |

[a]All $k = 1, 2, \ldots$, and all $r = 1, 2, \ldots$.

THEOREM 3.7. Let $Z_n$ represent the number of channel networks in a network pattern randomly selected from $\Omega_n$. Then the generating function $E\ t^{Z_n}$, the probability distribution $P\{Z_n = k\}$, and the rth factorial moment $E\ Z_n^{(r)}$ are as given in Table 3.2.

THEOREM 3.8. Put $Z = \lim_{n \to \infty} Z_n$. Then the generating function $E\ t^Z$, probability distribution $P\{Z = k\}$, and factorial moment $E\ Z^{(r)}$ are as given in Table 3.3.

It is convenient to interpret the six positive valued random variables identified in Tables 3.2 and 3.3 as shifted forms of commoner nonnegative valued random variables. The following two properties relate these six random variables to the family of generalized hypergeometric series random variables that have been studied extensively by Kemp (1968) and Dacey (1972b).

PROPERTY 3.1. The random variables $(X_n - 1)$, $(Y_n - 1)$, $(Z_n - 1)$, $(X - 1)$, and $(Y - 1)$ belong to the family of generalized hypergeometric series random variables (Kemp, 1968; Dacey, 1972b). The random variable $(Z_n - 1)$ has the generalized hypergeometric distribution type III (Patil and Joshi, 1968, p. 35; Johnson and Kotz, 1969, p. 159). The random variable $(X_n + 1)$ has the negative hypergeometric distribution (Johnson and Kotz, 1969, p. 157) or the inverse hypergeometric distribution (Patil and Joshi, 1968, p. 35). The random variables $(X - 1)$ and $(Y - 1)$ have the inverse Polya distribution (Patil and Joshi, 1968, p. 31-32; Johnson and Kotz, 1969, p. 232).

PROPERTY 3.2. The random variable $(Z - 1)$ has the negative binomial distribution with parameters 2 and ½ and generating function

$$E\ t^{Z-1} = \left\{\frac{\frac{1}{2}}{1 - \frac{1}{2}t}\right\}^2 = (2 - t)^{-2}$$

The next two properties identify urn models that generate the random variables X and Y. The first urn model provides for a derivation of link magnitude.

PROPERTY 3.3. The urn initially contains one black ball and one white ball. A ball is drawn from the urn and its color noted; then it is returned to the urn along with two additional balls of the same color. Balls are drawn from the urn until the first black ball is selected. Let X* denote the number of draws required to obtain one black ball. The link magnitude X is distributed as X*.

An interesting feature of this urn model, which has implications to the structure of channel networks, is that each failure to draw a white ball increases the probability that the next drawing is also not a white ball.

The next urn model provides a derivation of number of channel networks in a network pattern.

PROPERTY 3.4. The urn initially contains one black ball and one white ball. A ball is drawn from the urn, its color noted and returned to the urn. Balls are drawn from the urn until the black ball is drawn twice in a row. Let Z* represent the number of drawings required to obtain the black ball twice in a row. The number Z of channel networks in a network pattern is distributed as the random variable $Z^* - 1$.

These urn models display the chance component that generates the long tails of the probability laws for X and Z. One reflection of the high degree of skewness is given by the next property, which is followed by two properties that describe the shape of the probability distributions for $X_n$, $Y_n$, and $Z_n$.

PROPERTY 3.5. For $k > 0$

$$P\{X = k + 1 | X > k\} = \frac{1}{2(k + 1)}$$

and

$$P\{X = k + 1 | X > k\} = \frac{k + 1}{2(k + 2)}$$

PROPERTY 3.6. For fixed $n \geq 1$ and $0 \leq k \leq n - 1$, the probability $P\{X_n = n - k\}$ is at a minimum for $k = (n + 1)/4$, where [X] denotes the integer part of x. Moreover, if $(n + 1)/4$ is an integer, then $P\{X_n = n - k\} = P\{X_n = n - k + 1\}$. In particular, the graph of $P\{X_n = j\}$ is U-shaped with the minimum value occurring at one or two values for j, $P\{X_n = 1\} < P\{X_n = 2\}$ for $n \geq 2$, and $P\{X_n = n\} > P\{X_n = n - 1\}$ for $n \geq 4$.

PROPERTY 3.7. For $n \geq 2$ and $1 \leq k \leq n - 1$, $P\{Y_n = k\} \leq P\{Y_n = k + 1\}$ and $P\{Z_n = k\} \leq P\{Z_n = k + 1\}$, where the equality holds only for $k = 1$.

PROPERTY 3.8. $E\ Z_n \leq 3$ and $E\ Z = 3$.

This surprising result, established by Werner (1971), indicates that the expected number of channel networks in a network pattern never exceeds three.

The preceding random variables identify properties of complete channel networks and network patterns. Empirical applications frequently concentrate on the upper reaches of the drainage systems, and for the analysis of incomplete systems there is need for properties of segments of channel networks that are formed by links of small magnitude. The investigation of these properties utilizes the concept of DEFINITION 3.12.

DEFINITION 3.12. If a channel network $\lambda_n$ contains a magnitude k link $\ell_k$, let $\lambda_n(\ell_k)$ be the collection of links, forks, and nodal points that are formed by the k paths that extend from the downnode of $\ell_k$ to the upnodes of k magnitude $\ell$ links such that each of these paths includes the upnode of $\lambda_k$. The collection $\lambda_n(\ell_k)$ is called the subchannel network defined by $\lambda_n$ and $\ell_k$.

REMARK. If $\lambda_n(\ell_k)$ is a subchannel network, then $1 \leq k \leq n$. For $k = n$, $\lambda_n(\ell_n) = \lambda_n$ so that every channel network is a subchannel network. Furthermore, if the downnode of $\ell_k$ is treated as a source, then $\lambda_n(\ell_k)$ is a channel network of magnitude k.

Because each link in $\Lambda_n$ is a member of a channel network, it is possible to select from $\Lambda_n$ both a link and the channel network that contains the link. This link and channel network define a unique subchannel network. A link is randomly selected from this subchannel network, and its magnitude defines the three following random variables.

DEFINITION 3.13. Let $\ell$ represent a link randomly selected from the links that comprise the channel network in $\Lambda_n$, and let $\lambda_n(\ell)$ be the subchannel network defined by this link. Given that $\ell$ has magnitude k, let $S_n(k)$ be the magnitude of a link randomly selected from the links comprising $\lambda_n(\ell_k)$. If the magnitude of $\ell$ is not specified, let $S_n$ be the magnitude of a link ramdomly selected from the links comprising $\lambda_n(\ell)$. Put $S = \lim_{n \to \infty} S_n$.

Similar types of random variables also are defined with respect to the collection $\Omega_n$ of network patterns.

DEFINITION 3.14. Let $\ell$ denote a link randomly selected from the links comprising the channel networks in $\Omega_n$ and $\lambda(\ell,n)$ be the subchannel network defined by this link. Denote by $W_n$ the magnitude of a link randomly selected from $\lambda(\ell,n)$ and put $W = \lim_{n \to \infty} W_n$. Given that $\ell$ has magnitude k, let $W_n(k)$ denote the magnitude of a link randomly selected from $\lambda(\ell_k,n)$.

Some properties of these new random variables are identified next.

THEOREM 3.9. For all $n \geq 1$ and $1 \leq m \leq n$, $S_n(m) = W_n(m) = X_m$.

The identity $M_n(m) = X_m$ implies that a subchannel network has properties identical to those of a channel network of the same magnitude, with the consequence that a subchannel network may be studied without regard to the larger channel network to which it is spliced. The identity $W_n(m) = X_m$ implies that a subchannel network selected from a network pattern has properties that are identical to those of a channel network of the same magnitude. These two considerations imply that any subchannel network may be analyzed as a channel network independently of any larger structure of which it is a component.

The random variable $X_n$ identifies the magnitude of a link randomly selected from a subchannel network that is itself defined by a link randomly selected from the links comprising the channel networks in $\Lambda_n$. Because each link in $\Lambda_n$ defines a single subchannel network, $S_n$ may be interpreted as the magnitude of a link randomly selected from the links that comprise the collection of all subchannel networks that occur in $\Lambda_n$. Although properties of $S_n$ and $W_n$ are given by Dacey (1972c), properties are identified here only for S and W, which are of particular interest because they represent the magnitude of links randomly selected from the links that comprise the subchannel networks that occur in $\Lambda$ and $\Omega$. Because it has been established that X = Z, the identity of the following theorem is not surprising.

THEOREM 3.10. S = W.

THEOREM 3.11. The probability distribution and generating function of the random variable S are

$$P\{S = k\} = 2^{-4(k-\frac{1}{2})} \frac{1}{k} \binom{2k-2}{k-1}^2 \qquad k = 1, 2, \ldots$$

and

$$E\, t^S = \frac{\pi}{4}\, {}_2F_1[\tfrac{1}{2}, \tfrac{1}{2}; 2, t]$$

The moments of S do not exist.

PROPERTY 3.9. The random variable (S - 1) has the inverse Polya distribution (Patil and Joshi, 1968, p. 31-32; Johnson and Kotz, 1969, p. 323).

PROPERTY 3.10. For all $K \geq 1$

$$P\{S = k\} = 2^{-(2k-1)} \binom{2k-1}{k-1} P\{X = k\} < P\{X = k\}$$

and

$$P\{S = k+1\} = \frac{(2k-1)^2}{4(k+1)k^2} P\{S = k\} < P\{S = k\}$$

## CLASSIFICATION OF LINKS

This classification of links uses properties that reflect the position of a link in a channel network. The class to which a link $\ell$ is assigned depends upon its magnitude relative to the magnitude of the links, if any, that are members of the forks at the upnode and downnode of $\ell$. These link magnitudes define nine classes of links. The probability that a link belongs to each class is obtained for a link randomly selected from the links comprising the topologically distinct magnitude n channel networks in $\Lambda_n$. This classification scheme, and the resulting probability distribution, are slight modifications of those proposed by Mock (1971). This classification scheme uses, without change, the definitions and structure developed in, particularly, the first section. The specification of branches by Definition 3.1 is, however, extended by Definition 3.15.

DEFINITION 3.15. A fork is formed by the coincidence of nodes of three distinct links: the downnodes of two distinct links, say, $\ell_1$ and $\ell_2$, and the upnode of a third link, say $\ell_3$. Links $\ell_1$ and $\ell_2$ are oriented with respect to $\ell_3$, and are called the left branch and right branch of link $\ell_3$; they also are called up branches of link $\ell_3$. Link $\ell_1$ is called the down branch of link $\ell_2$ and, conversely, $\ell_2$ is called the down branch of link $\ell_1$. Figure 3.5 illustrates this terminology.

The collection $\Lambda_n$ contains N(n) magnitude n channel networks that are distinguishable, and each such network contains (2n - 1) links. Moreover, each of the (2n - 1)N(n) links in $\Lambda_n$ is a member of a channel network. Hence, it is possible to select simultaneously from $\Lambda_n$ a link and the channel network of which it is a member. The classification of a link depends upon its position in a channel network.

DEFINITION 3.16. Let $X_n$ denote a link randomly selected from the links in $\Lambda_n$ and let $\lambda_n$ denote the channel network in $\Lambda_n$ that contains $X_n$. If $X_n$ is a link of the channel network $\lambda_n$, this is written as $\{X_n \in \lambda_n\}$. If $X_n$ has magnitude k, this event is denoted by $\{X_n = k\}$.

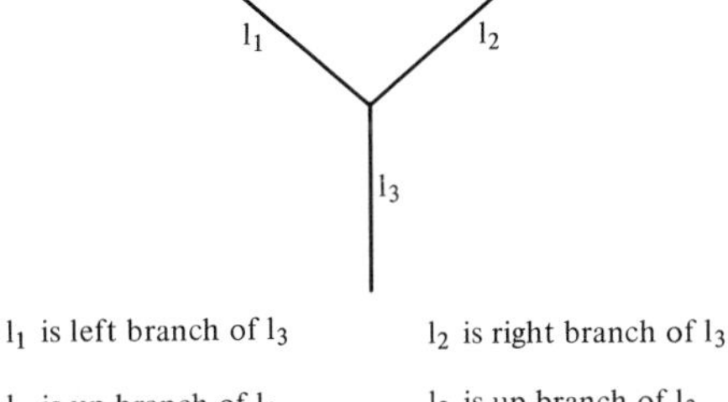


FIGURE 3.5 Example of terms identifying types of branches.

If $X_n$ has the class property A, i.e., belongs to class A, this event is denoted by $\{X_n \in A\}$.

DEFINITION 3.17. If $\{X_n > 1\}$ occurs, let $U_n$ and $U_n'$ denote the magnitudes of the left branch and the right branch of $X_n$. Also, if $\{X_n < n\}$ occurs, let $V_n$ denote the magnitude of the down branch of $X_n$.

Notice that when $\{X_n = 1\}$ occurs, the up branches of $X_n$ do not exist so that $U_n$ and $U_n'$ are not defined. Likewise, when $\{X_n = n\}$ occurs, the down branch of $X_n$ does not exist so that $V_n$ is not defined.

The magnitude of a link may be expressed in terms of the magnitude of its up branches.

PROPERTY 3.11. If $X_n$ is an exterior link so that the up branches of $X_n$ do not exist, then $\{X_n = 1\}$ occurs. If $X_n$ is not an exterior link, so that $U_n$ and $U_n'$ are defined, then $\{X_n = U_n + U_n'\}$ occurs.

The classification of the link $X_n$ depends upon its magnitude and the magnitude of the links that form its branches. The classification in terms of its up branches only reflects whether $U_n$ and $U_n'$ have the same or different magnitudes. Because of this property, it does not matter whether $U_n$ is interpreted as the magnitude of the left or right branch of $X_n$.

Each link $X_n$ is assigned to one of nine classes that reflects its upnode and downnode properties. The three upnode classes depend upon the relative magnitudes of $X_n$ and $U_n$, and the three downnode classes depend upon the relative magnitudes of $X_n$ and $V_n$.

DEFINITION 3.18. The three upnode classes are A, B, and C.

$\{X_n \in A\}$ occurs iff $\{X_n > 1\}$ and $\{X_n \neq 2U_n\}$ both occur.
$\{X_n \in B\}$ occurs iff $\{X_n > 1\}$ and $\{X_n = 2U_n\}$ both occur.
$\{X_n \in C\}$ occurs iff $\{X_n = 1\}$ occurs.

DEFINITION 3.19. The three downnode classes are D, E, and F.

$\{X_n \in D\}$ occurs iff $\{X_n < n\}$ and $\{X_n \geq V_n\}$ both occur.
$\{X_n \in E\}$ occurs iff $\{X_n < n\}$ and $\{X_n < V_n\}$ both occur.
$\{X_n \in F\}$ occurs iff $\{X_n = n\}$ occurs.

Notice that if $\{X_n \in C\}$, then $X_n$ is an exterior link. Otherwise, it is an interior link. If $\{X_n \in F\}$, then $X_n$ is a terminal link.

The class of $X_n$ is determined by the union of its upnode and downnode classes. For example, the event $\{X_n \in AD\}$ occurs if, and only if, $\{X_n \in A\}$ and $\{X_n \in D\}$ both occur. The nine classes of links are identified in Table 3.4. The name for each class is suggestive of its dominant property. For

TABLE 3.4 Link classes

| The Nine Link Classes | | Mock's Classification[a] |
|---|---|---|
| $\{X_n \in DC\} = \{1 = X_n < n, X_n \geq V_n\}$ | Source | $\{X_n = 1, X_n = V_n\} \equiv D$ |
| $\{X_n \in EC\} = \{1 = X_n < n, X_n < V_n\}$ | Tributary source | $\{X_n = 1, X_n < V_n\} \equiv ED$ |
| $\{X_n \in DB\} = \{1 < X_n < n, X_n = 2U_n, X_n \geq V_n\}$ | Bifurcating | $\{X_n > 1, X_n = 2U_n, X_n > V_n\} \equiv B$ |
| $\{X_n \in EB\} = \{1 < X_n < n, X_n = 2U_n, X_n < V_n\}$ | Tributary bifurcating | $\{X_n > 1, X_n = 2U_n, X_n \leq V_n\} \equiv EB$ |
| $\{X_n \in DA\} = \{1 < X_n < n, X_n \neq 2U_n, X_n \geq V_n\}$ | Cis and trans | $\{X_n > 1, X_n \neq 2U_n, X_n > V_n\} \equiv CE$ |
| $\{X_n \in EA\} = \{1 < X_n < n, X_n \neq 2U_n, X_n < V_n\}$ | Tributary | $\{X_n > 1, X_n \neq 2U_n, X_n \leq V_n\} \equiv E$ |
| $\{X_n \in FB\} = \{1 < X_n = n, X_n = 2U_n\}$ | Terminal bifurcating | Not defined |
| $\{X_n \in FA\} = \{1 < X_n = n, X_n \neq 2U_n\}$ | Terminal | Not defined |
| $\{X_n \in FC\} = \{n = 1\}$ | Isolated | Not defined |

[a]From Mock (1971).

comparison, the six types of links identified by Mock are expressed in the same notation. Whereas the new classification has three classes of terminal links that are not incorporated in Mock's scheme, these additions are not critical because a channel network has only one terminal link. The main distinction is that the new scheme considers $X_n$ in relation to $V_n$ as "less than" or "equal to or greater than," whereas Mock uses four types of relations. The gain from this minor alteration is that the exterior and interior links are classified on a consistent basis. In contrast, Mock classifies exterior links by a "less than" and "equal to" dichotomy and interior links by a "less than or equal to" and "greater than" dichotomy.

THEOREM 3.12. Table 3.5 gives the probability that $\{X_n \in \alpha\}$ occurs for $\alpha$ replaced by A, B, C, D, E, and F.

TABLE 3.5 Probabilities for upnode and downnode classes

Definition

(1) $$J(n) = \tfrac{1}{2} \sum_{1 \leq k \leq \frac{1}{2}n} \pi(k,n)\pi(k,n-k)$$

(2) $$K(n) = \left[ \sum_{1 \leq k \leq \frac{1}{2}n} \sum_{i=1}^{k} + \sum_{\frac{1}{2}n < k < n} \sum_{i=1}^{n-k} \right] \pi(i, n-k)\pi(k,n)$$

Upnode Class Probabilities

(3) $$P\{Y_n \in A\} = \frac{n-1}{2n-1} - J(n) \qquad n > 1$$

(4) $$P\{Y_n \in B\} = J(n) \qquad n > 1$$

(5) $$P\{Y_n \in C\} = \frac{n}{2n-1}$$

Downnode Class Probabilities

(6) $$P\{Y_n \in D\} = K(n) \qquad n > 1$$

(7) $$P\{Y_n \in E\} = \frac{2n}{2n-1} - K(n) \qquad n > 1$$

(8) $$P\{Y_n \in F\} = \frac{1}{2n-1}$$

THEOREM 3.13. Table 3.6 gives the probability that link $X_n$ randomly selected from the links in $\Lambda_n$ is a member of each of the nine classes of links.

THEOREM 3.14. Put $X = \lim_{n \to \infty} X_n$. Table 3.7 gives the probability that $\{X \in \alpha\}$ occurs for $\alpha$ replaced by A, B, C, D, E, and F.

TABLE 3.6 Probabilities for link classes

Definitions

J(n) and K(n): see Table 3.5

(1) $$G(n) = \pi(\tfrac{1}{2}n,\ n-1)\pi(n,n) = \tfrac{1}{4}\pi^{-\frac{1}{2}} \frac{\Gamma(\frac{1}{2}n-\frac{1}{2})\Gamma(\frac{1}{2}n-\frac{1}{2})\Gamma(n)}{\Gamma(\frac{1}{2}n)\Gamma(\frac{1}{2}n+1)\Gamma(n+\frac{1}{2})} \qquad n \text{ even}$$
$$= 0 \qquad n \text{ odd}$$

(2) $$H(n) = \left[\sum_{1 \leq k < \frac{1}{4}n} \sum_{i=1}^{2k} + \sum_{\frac{1}{4}n \leq k \leq \frac{1}{2}(n-1)} \sum_{i=1}^{n-2k}\right] \cdot \pi(i, n-2k)\pi(k, 2k-1)\pi(2k, n)$$

Class Probabilities for n = 1

(3) $P\{X_1 \in CF\} = 1$

Class Probabilities for n > 1

(4) $P\{X_n \in AD\} = \dfrac{1}{2n-1} - G(n) \qquad n \geq 3$

(5) $P\{X_n \in BD\} = G(n) \qquad n \geq 2$

(6) $P\{X_n \in CD\} = 0 \qquad n \geq 2$

(7) $P\{X_n \in AE\} = \dfrac{3n^2 - 8n + 6}{(2n-1)(2n-3)} + G(n) + H(n) - J(n) - K(n) \qquad n \geq 7$

(8) $P\{X_n \in BE\} = J(n) - G(n) - H(n) \qquad n \geq 5$

(9) $P\{X_n \in CE\} = \dfrac{n(n-2)}{(2n-1)(2n-3)} \qquad n \geq 3$

(10) $P\{X_n \in AF\} = K(n) - H(n) - \dfrac{n(n-1)}{(2n-1)(2n-3)} \qquad n \geq 4$

(11) $P\{X_n \in BF\} = H(n) \qquad n \geq 3$

(12) $P\{X_n \in CF\} = \dfrac{n(n-1)}{(2n-1)(2n-3)} \qquad n \geq 2$

TABLE 3.7 Limiting probabilities for link classes

| | $P\{X \in \alpha\beta\}$ | | | | | |
|---|---|---|---|---|---|---|
| $\beta/\alpha$ | A | | B | | C | Sum |
| D | 0.29976 | | 0.08686 | | 0.25000 | 0.63662 $(=2/\pi)$ |
| E | 0.06362 | | 0.04976 | | 0.25000 | 0.36338 $(=1 - 2/\pi)$ |
| F | 0 | | 0 | | 0 | 0 |
| Sum | 0.36338 | $=1-(2/\pi)$ | 0.13662 | $=(2/\pi)-(1/2)$ | 0.50000 | 1.00000 |

Mock's Classes

| Class | Probability | Class | Probability |
|---|---|---|---|
| Cis and trans | 0.29251 | Tributary | 0.07087 |
| Bifurcating | 0.07087 | Tributary bifurcating | 0.06575 |
| Source | 0.25000 | Tributary source | 0.25000 |

THEOREM 3.15. Put $X = \lim_{n \to \infty} X_n$. Table 3.7 gives the probability that $\{X \in \alpha\beta\}$ occurs for $\alpha$ replaced by A, B, and C and $\beta$ replaced by D, E, and F.

THEOREM 3.16. The bottom part of Table 3.7 gives the probability that $\{X \in \alpha\}$ occurs when $\alpha$ is replaced by each of the six classes in Mock's classification, as identified in Table 3.4. Note that four of these values differ by 0.00001 from the estimates obtained by Mock.

## EMPIRICAL IMPLICATIONS

A large number of empirical studies have established that the model of link magnitudes for channel networks has a high level of conformity with magnitude numbers of streams in regions that lack strong geologic controls. Also, the more limited empirical evidence given by Mock indicates good agreement between observed and theoretical proportions of link classes. In contrast, it has been difficult to identify a physical interpretation for the model of network patterns. Dacey's (1972a) suggestion that it pertains to the drainage system within an arbitrary region, such as a map quadrangle (Figure 3.6), is contradicted by the empirical evidence. A major obstacle to interpretation of the model is that the expected number of channel networks in a network pattern does not exceed three, and it is difficult to identify the regional or spatial structure

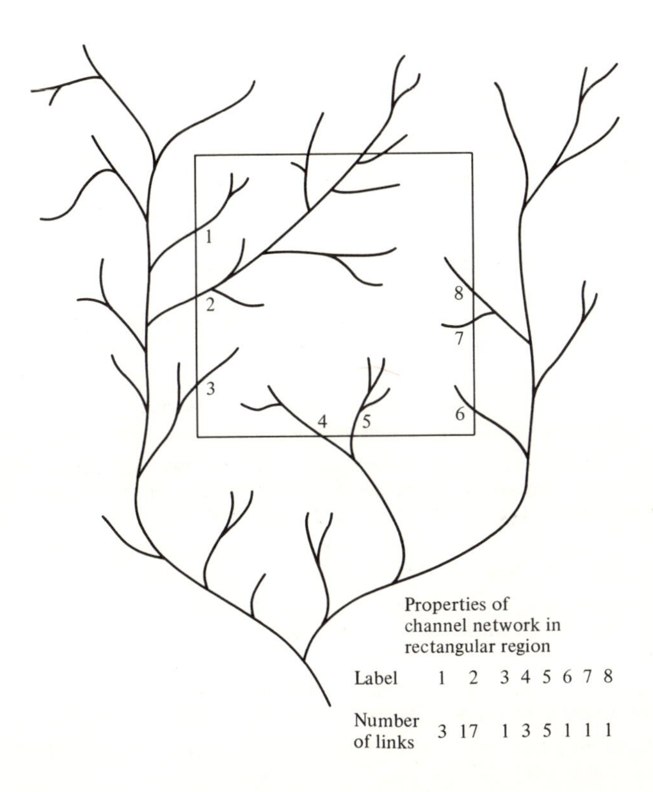


FIGURE 3.6 Illustration of possible interpretation of model of network patterns; ordering of component channel networks is determined by ordered arrangement of exit points.

of drainage systems that have this property. One possibility for this lack of empirical verifiability is that the model of network patterns is based only on the concept of topologic randomness and is, accordingly, incomplete in that it neglects basic properties of the spatial configuration of drainage systems. If so, the need now is to integrate the structure of channel networks into models of network patterns that incorporate both topological randomness and spatial randomness.

## ACKNOWLEDGMENT

The support of the National Science Foundation, Grant GS-2967, is gratefully acknowledged.

## REFERENCES

Dacey, M. F., 1971, Probability distribution of number of networks in topologically random network patterns: Water Resources Res., v. 7, no. 6, p. 1652-1657.

Dacey, M. F., 1972a, Some properties of link magnitude for channel networks and network patterns: Water Resources Res., v. 8, no. 4, p. 1106-1111.

Dacey, M. F., 1972b, A family of discrete probability distributions defined by the generalized hypergeometric series: Sankhya, Ser. B, v. 34, no. 2, p. 243-250.

Dacey, M. F., 1972c, A review of number properties of channel networks and network patterns: Research Rept. No. 60, Dept. Geography, Northwestern Univ., Evanston, Illinois, 169 p.

Erdelyi, A., ed., 1953, Higher transcendental functions (The Bateman Manuscript Project, Vol. I): McGraw-Hill Book Co., New York, 302 p.

Horton, R. E., 1945, Erosional development of streams and their drainage basins: hydrophysical approach to quantitative morphology: Geol. Soc. America Bull., v. 56, no. 3, p. 275-370.

Johnson, N. L., and Kotz, S., 1969, Discrete distributions: Houghton-Mifflin, Boston, Massachusetts, 328 p.

Kemp, A. W., 1968, A wide class of discrete distributions and the associated differential equations: Sankhya, Ser. A, v. 30, no. 4, p. 401-410.

Mock, S. J., 1971, A classification of channel links in stream networks: Water Resources Res., v. 7, no. 6, p. 1558-1566.

Morisawa, M., 1968, Streams: their dynamics and morphology: McGraw-Hill Book Co., New York, 175 p.

Patil, G. P., and Joshi, S. W., 1968, A dictionary and bibliography of discrete distributions: Oliver and Boyd, Edinburgh, 268 p.

Shreve, R. L., 1966, Statistical law of stream numbers: Jour. Geology, v. 74, no. 1, p. 17-37.

Shreve, R. L., 1967, Infinite topologically random channel networks: Jour. Geology, v. 75, no. 2, p. 178-186.

Shreve, R. L., 1969, Stream lengths and basin areas in topologically random channel networks: Jour. Geology, v. 77, no. 4, p. 397-414.

Strahler, A. N., 1952, Hypsogeometric (area-altitude) analysis of erosional topography: Geol. Soc. America Bull., v. 63, no. 11, p. 1117-1142.

Strahler, A. N., 1964, Quantitative geomorphology of drainage and channel networks, <u>in</u> Handbook of applied hydrology: McGraw-Hill Book Co., New York, various pagings.

Werner, C., 1971, Expected number and magnitudes of stream networks in random drainage patterns: Assoc. Am. Geographers Proc., v. 3, p. 181-185.

# Probabilistic Modeling in Geology

W. C. Krumbein

ABSTRACT

Simulation studies of geologic phenomena with probabilistic models based on independent-events mechanisms, Markov processes, and Poisson processes, are increasingly common. Arbitrary selection of a given model because its simulation output resembles the system under study may be misleading or, at best, may shed only minimal light on the physical controls that actually determine the behavior of the system.

As in other sciences, stochastic process modeling in geology is a formal procedure that normally requires conceptualization of the process followed by identification and mathematical expression of the elements that control the process. This procedure (in which the conceptualization stage may be mainly qualitative) forces the geologist to be specific in his statements concerning the essential elements in his study and the relations among them. These elements then are combined in a "reasonable" probabilistic setting, which may have one or more deterministic controls.

The advantages of this formal sequence are that they normally lead to an initial model that is highly flexible and lends itself well to sequential modeling procedures. The steps involved are illustrated with examples from paleontology and sedimentation.

## INTRODUCTION

Among recent developments in the classical fields of earth science is the increasing use of probabilistic models for describing and analyzing natural phenomena. It is customary in the present early stages of this development to draw mainly upon conventional probabilistic mechanisms, such as random walks, Markov chains, and independent-events models, inasmuch as their mathematical properties are well known, and observational data for testing the models are available or can readily be obtained.

It becomes evident as experience increases that currently these models are used mainly as descriptive devices. That is, they support the inference that probabilistic elements are present in many geologic phenomena, but they do not directly "explain" the mechanism in terms of underlying physical controls. The models, nevertheless, are important in establishing a framework for subsequent more critical analysis.

A Markov transition probability matrix, for example, is an excellent device for summarizing stratigraphic sequences. It gives a good estimate of the relative probabilities with which lithologic units follow each other in a stratigraphic section; moreover, major depositional cycles can be discerned and even subcycles within them identified. This information is valuable for reconstructing the succession of environments in basins of sedimentation, which, in turn, suggest combinations of physical and chemical conditions that may succeed each other.

Perhaps the greatest value of the models as presently used is their strong implication that some random elements do enter many geologic phenomena. This result can be used as a starting point for developing probabilistic models specifically designed for particular geologic processes. The main purpose of this paper is to describe this emerging pattern of model building. We start our discussion with a review of standard models already in use.

## CONVENTIONAL PROBABILISTIC MODELS IN GEOLOGY

The most widely used model in the context of this paper is the first-order discrete-time discrete-state Markov chain. This particular model lends itself well to stratigraphic analysis, in which the states of the system are defined as the lithologic components (sandstone, shale, limestone, etc.) that occur in the sections being studied. For these situations the transition probabilities $p_{ij}$ are estimated. Simulations from the matrices yield synthetic sections similar to those seen in nature, especially for cyclical deposits.

Similarity of visual appearance between Markovian simulation output and natural stratigraphic sections is not an adequate test of the model. Common practice now is to test two aspects of the observed stratigraphic input data before accepting the model. The first is the presence of the Markov property, i.e., that the state of the system at time $t_{n+1}$ depends only on its state at $t_n$, or on longer memories, as discussed by Schwarzacher (1967). The second test is whether the population densities of the natural lithologic unit thicknesses are geometric, with parameter $(1 - p_{ii})$, where $p_{ii}$ is the probability that a given state succeeds itself on successive draws.

When the first of these conditions is satisfied and the second is not, an embedded Markov chain with $p_{ii}$ identically zero is appropriate (Krumbein and Dacey, 1969). This variant removes the requirement of a geometric waiting time, and permits use of any suitable thickness density. If the opposite condition holds, namely, that the Markov property is absent but the input data do have the geometric distribution, then an independent-events model is appropriate. For this latter model, the state of the system at $t_{n+1}$ is independent of its state at $t_n$.

Shreve (1966, 1967) applied an independent-events model to stream drainage networks by postulating that all possible topological configurations in stream basins of given magnitude are equally likely in the absence of geologic controls. This leads to a uniform density for the topological classes involved, easily verified for low-magnitude basins, and somewhat less directly (Smart, 1969) for larger magnitude basins. One geometric distribution related to topology is directly implied in this model: that clusters of size k (k = 1, 2, ...) of equal-magnitude contiguous basins with the same topological configuration are distributed geometrically with a parameter predictable from the corresponding uniform distribution. This was tested with observed data on basin clusters of magnitudes 4 and 5 (Krumbein, 1970), which agree with expectations.

Random-walk models of the type introduced by Leopold and Langbein (1962) for stream profiles and stream network generation can be converted to independent-events models or to Markov chains, depending on the structure of the models. The stream profile, for example, can be expressed as a Markov chain with absorbing terminal state. Smart (1968) postulated a random stream-generating process that led directly to a geometric distribution of exterior link lengths. This provides a reasonable initial process; it calls for subsequent channel adjustments that produce Gamma or Gammalike link-length distributions that Shreve (1969) showed to agree more favorably with observed data on natural streams. In addition, a first-order Markov dependency tends to develop along main channels in the downstream successive order of tributary entrance (James and Krumbein, 1969).

These details are recounted here for two reasons: They emphasize both the importance of the initial model (i.e., Shreve's postulate of topological randomness) and the almost inevitable subsequent introduction of real but relatively second-order qualifications as more data accumulate. This leads to the first point of emphasis of this paper - that modeling is essentially a sequential process.

## SOME LIMITATIONS OF CONVENTIONAL MODELS

Probabilistic models, similar to their deterministic counterparts, must ultimately include the physical or chemical processes that underlie the phenomena, in order to become fully analytical. Present descriptive uses of the models, as mentioned, may fail to discriminate among the several choices available for studying a given process. Howard (1971) recently emphasized that stream-network simulations from a variety of random-walk models yield output similar in appearance, which agrees about equally well with properties of the input data. As Howard pointed out, this suggests that these models may be too general for their purpose. Similarly, Smart (1971) urged detailed study of proposed models to identify the circumstances in which they fail to conform with nature, as well as those aspects in which they do.

This is, in fact, normal procedure with deterministic models, where "equation error" (Krumbein and Graybill, 1965, p. 17) enters the model if predictions extend beyond the point where the model applies. Stokes' law of settling velocities of spheres is an example in point, where equation error enters strongly when viscous forces are no longer dominant in controlling the motion.

Despite present limitations in conventional probabilistic models, one important contribution of their diffusivity is that they demonstrate that more than a single control may give rise to the same response. Circumstances occur (as in bifurcation of stream channels) in which the bifurcation response can be produced by more than one cause. This necessarily moves the framework of analysis and prediction of such events from a unique one-to-one cause-and-effect couplet (where predictions can be made with probability $p = 1$) to the domain of a sample space in which predictions depend on the relative probabilities of a given response produced by more than a single control. Thus, as Watson (1969) points out, prediction of specific outcomes shifts to $0 < p < 1$.

## SUBSTANTIVE STATISTICAL MODELS

Exploration of new avenues opened by recognition of probabilistic elements in geologic phenomena can be pursued along several paths. One method is to use a deterministic core and to build into it one or more random effects. For such a model to have physical meaning, the random effects must have some relation to the process under study. Where diffusion is the main physical process, the strategy used by McEwen (1950) can be effective. He developed a statistical equivalent of diffusion by deriving a random-walk model, the parameters of which then were equated to the corresponding constants in the differential equation for diffusion. This yields a stochastic counterpart of the diffusion equation that contains a variance term with direct physical meaning. The model was applied successfully to observations of drift bottles in the sea, as well as to temperature and salinity in stagnant areas.

A somewhat different approach, leading to a "substantive statistical model," was introduced by James (1967). The process is first expressed as a set of discrete qualitative statements that describe the conceptual model, i.e., a mental picture of what is going on. Each statement then is transformed into a formal algebraic or probability definition or equation. The qualitative statements are erased, as it were, and the purely mathematical relations among the equations are developed to yield an initial model tailored to the specific process under study. An advantage of this type of modeling is that the parameters and coefficients have direct physical meaning, and hence the model itself specifies the types of observational data needed to test its own adequacy.

The initial model, as just described, is seldom sufficiently inclusive or general to cover the whole of the process under study. Its importance arises from the fact that it presents a particular (and perhaps new) way of looking at a phenomenon, and hence it is the starting point for further studies.

## THE INITIAL MODEL

The foregoing remarks introduce the second point of emphasis in this paper - that the initial model sets the framework for its own sequential modification. As such, it deserves detailed comment. If an initial model is far from the mark as tested by observational data, it is normally discarded, and thus starts and ends a modeling sequence with $n = 1$ step. We assume here that the model is sufficiently good so that it survives its initial tests.

Model survival depends in part on how the initial qualitative statements are viewed. Coleman (1964, p. 522) points out that qualitative statements about a given phenomenon can be used in either of two ways for model building. One is to view the statements as postulates for the model; the other is to consider them as consequences arising from the process involved in the model. Coleman cites examples of both approaches and recommends the second one.

Whether this distinction will be as clearly cut in geologic modeling is not evident at this writing, in part because the original qualitative statements are converted to quantitative mathematical or probability equivalents. The point raised by Coleman does suggest, however, that the foundation of qualitative statements be examined critically, and perhaps divided into two groups, one leading to model postulates and the other to consequences of the model. In one sense this may imply that the statements can be divided into those that define process elements as opposed to those that are responses to the process. In this situation the second group of statements are themselves tests of the model.

## SCHEME FOR SEQUENTIAL MODELING

The two statements on the sequential nature of modeling and the importance of the initial model can be used to develop a flowsheet for model building. Table 4.1, modified from earlier attempts along the same line (cf. Krumbein and LaMonica, 1966, p. 347), shows what I believe to be an emerging pattern. The table is not intended to formulate any rules about building models, but rather to suggest that a logical succession of steps can be discerned.

As in virtually all geologic studies, a background of field experience, discussions with colleagues, and literature scanning is assumed on the part of the researcher as a preliminary to formal model building. This is indicated as step 1 in Table 4.1. Step 2, in many situations, concurrent with the first, is the development of a conceptual model, which as stated may be mainly qualitative or diagrammatic, although not necessarily so. Formalization begins at step 3 when the conceptual model is arranged as a series of specific statements related to the process or its responses. Presumably the process statements are used as postulates, but see Coleman's distinction in the previous section.

Step 4 in Table 4.1 is a pivotal point in the modeling procedure. It is here that the statements are converted to mathematical expressions by any of a variety of strategies. Each statement can be converted to a specific probability distribution or to a known stochastic subprocess that seems appropriate to the study. Alternatively, each statement can be converted to a probability expression without commitment at this step to any specific density. Other approaches include the use of a deterministic core, as in the McEwen example cited earlier, or to an explicitly deterministic expression in the manner used by Briggs and Pollack (1967) in their Michigan Basin evaporite model.

TABLE 4.1 Generalized flowsheet for geologic modeling

1. Identification and selection of a particular problem for study.
2. Development of a conceptual model (i.e., a mental picture) of the process and its anticipated responses. This may be expressed as a diagram, a process-response flowchart, a compact statement in words, or even as an initial arrangement of the mathematical aspects of the problem.
3. The conceptual model is reduced to a series of discrete statements, each related to one or more of the specific process elements in the conceptual model. Response elements may be put aside for later use in testing the model.
4. Each process statement in step 3 is formalized as a specific algebraic, probabilistic, or other mathematical expression that is isomorphic with the corresponding statement.
5. The original statements in step 3 are put aside, and the purely mathematical relations among the symbolic expressions of step 4 are condensed into one or more equations (differential or otherwise) that can be solved by analytical or numerical methods. This is the initial model.
6. Solution of the equations in step 5 with such boundary conditions, theoretical, or empirical values of the constants or parameters as may already be available, or whose structure and dimensions are adequate for translation into quantities measurable by observation.
7. Testing the model with observational data mainly specified by the initial model itself, as well as by verifying that the output conforms to the response statements in step 3.
8. Decision to reject, accept, or modify the initial model. Modification may start at any of the preceding steps, depending on the results of step 7. The need for modification may be noticeable at the outset, or may become apparent as more data are made available for the model.

Decisions at this stage mainly determine whether the model is to be deterministic or dominantly probabilistic. The Briggs and Pollack model, for example, is classified by its authors as purely deterministic. Emphasis in this paper is on probabilistic models, although the flowchart can just as readily include deterministic modeling. My choice is one of convenience, in that the probabilistic approach seems to me to offer more versatility.

The dominantly mathematical step is step 5 in the sequence of Table 4.1. It is here that the mathematical and probability theory background and ingenuity of the researcher come into full play. Ideally, all the expressions can be reduced to a single mathematical equation or to a set of simultaneous equations that serve as the initial model. Success at this stage depends on whether the equations can be solved by classic or numerical methods in step 6 to yield an expression in which at least some terms can be estimated from direct observations. Model testing - with resultant rejection, modification, or acceptance - are steps 7 and 8 of the flowsheet.

It is not unusual in this model-building scheme to discover that some of the critical observational data required to test the model are either not available or may not be obtainable in terms of present measurement capabilities. In this situation the researcher may: (1) publish his model and propose it as one that will prove successful if the necessary types of data become available; (2) reduce his model to a less all-embracing process; or (3) extend the analysis to include the types of data presently available for observation.

One of the most stimulating aspects of this new tendency in geologic modeling is that it forces the geologist to be specific in his statements concerning the essential elements in his study and the relations among them. This does not by any means exclude the use of standard statistical models, such as stepwise regression, trend analysis, factor analysis, etc., but these now take their position as "search models" in helping the geologist identify the more important variates to be included in his conceptual model. In short, conventional statistical analysis in this framework plays its strongest role in setting the stage for the first step in Table 4.1.

The principal differences between the modeling procedures in Table 4.1 and conventional quantitative studies in classic geology deserve comment. This is deferred to the Discussion section, after two examples are cited.

## PROBABILISTIC MODELING IN PALEOECOLOGY AND SEDIMENT TRANSPORT

Through the courtesy of Douglas Lorenz (1973) this example illustrates the development of a probabilistic model for observed occurrences of brachiopod patches in the Eden Shale (Ordovician) of north-central Kentucky. Field study showed a series of discrete lenticular beds of highly fossiliferous limestone scattered laterally and vertically through the shale. Almost all of the data were obtained from outcrops and roadcuts, which display the patches in cross section only. In a single instance a number of patches were seen as roughly circular lenses parallel to the bedding in planar view. The aspect of Lorenz's work described here concerns processes that control the size distribution of the limestone patches, taking into account the limitation of observations to outcrop data. The full

study is considerably broader than this, and the detailed mathematical steps are omitted, inasmuch as they are fully developed in Lorenz's forthcoming publication. The main concern here is the manner of implementing the statements in the conceptual model with probabilistic counterparts, as in steps 3 and 4 of Table 4.1.

Lorenz's conceptual model and the probability assumptions made for each statement can be succinctly stated as follows: The brachiopod patches are initiated from previous colonies by spatfall on favorable mud-bottom sites, assumed to be Poisson distributed over the area involved. The patches grow by a random-walk diffusion process that results in radial growth proportional to the age of the patch. This postulates a birth-death process directly adopted from Bharucha-Reid (1960, p. 88). The episodes of mud invasion, presumably controlled by changes in source areas and sediment-dispersal patterns, are taken to be random in time and space. The probability of a given patch being smothered by these random influxes of mud is considered to be constant through time.

By using probability theory to gain an initial model for this single aspect of his study, Lorenz showed that the population density function of patch radius R of the entombed patches is exponential: $f(R) = \alpha e^{-\alpha R}$, where $\alpha$ has the form $\alpha = \phi/(cd^2)^{1/2}$. Here $\phi$ is the rate of current shifting, c is Bharucha-Reid's intrinsic rate of population growth, and $d^2$ is the mean square displacement of the brachiopods per generation.

At this point a problem arises in testing the initial model for the frequency distribution of patch radii. As mentioned, the observational data are limited to outcrops that display the patches in cross section only. Moreover, some patches are longer than their outcrops are wide, thus yielding data obviously truncated at the upper end. To handle this problem, Lorenz invoked Sievert's (1930) probability distribution of linear intercepts produced by random sectioning of circles. In addition, it was necessary to express patch length as the observed length of patches with both ends exposed, divided by the outcrop length. Sievert's distribution, modified for this new definition of intercept length, then was truncated at the point where this ratio is 1.0. The observed distribution of patch length agreed well with prediction by conventional $\chi^2$ tests.

This particular example is treated in such detail in order to illustrate two points: (1) Lorenz's logical step-by-step development of his model, and (2) his reliance on probability distributions and stochastic expressions available in the literature. It should be mentioned that in examining the spatial distribution of the patches, his assumption of

Poisson randomness was weak, so that his initial model, with respect to other patch attributes, provides a reasonably satisfactory first stage of sequential modeling.

A second example of probabilistic modeling, in which the mathematical implementation of the statements in the conceptual model avoid commitment to any specific population densities, is afforded by James' (1970) model for movement of sand along a coastline by wave and current action. The problem involved is to develop a field method for measurement of the short-term volume littoral drift rate. This is of basic importance in coastal engineering, where the design and location of shore structures depend heavily on such data.

The process of littoral drift involves both transport and burial of the material on the nearshore bottom. Radioactive tracer sand can be used to measure the areal rate of tracer transport along the coast, but there is no obvious method of measuring the depth of transport, which is needed to convert the areal rate to the required volume transport rate.

James designed a conceptual two-dimensional sediment-transport model that includes only direction of transport and depth of burial. As a wave crest passes over a point, particles are lifted from the bed, transported, and redeposited. When waves approach the shore at an angle, a net long-shore current is generated for carrying the particles downbeach. Inasmuch as wave height and period are random variables, the depth to which material is eroded differs with each wave.

Accordingly, James started with the concept of a point on the seafloor, the elevation of which is considered to change instantaneously through a series of erosional and depositional increments. The eroded particles are assumed to mix in transport so that the depth below the sediment-water interface at which they are redeposited is independent of the depth from which they were eroded. The tracer particles are emplaced on the seafloor at time zero. They are eroded, transported, and redeposited along with untagged bed material until an equilibrium state is reached in which the concentration of tracer particles with depth no longer changes.

These conceptual statements were first implemented in step 4 of Table 4.1 by defining the erosional increment of the ith surge as $\varepsilon_i$ and the thickness of the subsequent depositional increment as $\delta_i$, both considered as random variables with common expectation $\mu$. A series of probability statements then is introduced, the first of which is that the probability of a particle being buried between depth z and (z + dz) is $f_{i+1}(z)dz$. From this and additional probability statements, James proceeds by combinatorial analysis to the equilibrium state $f_{eq}(z)$, when the condition $f_{i+1}(z) = f_i(z)$ is satisfied. This development requires no knowledge or assumptions regarding the specific probability density functions of any of his conceptual statements.

From this point on, by an interplay of definitions and combinations, James' analysis leads to a special situation of the equilibrium distribution of burial depths, $f_{eq}(0) = (Pm_{eq}/\mu)$, where $Pm_{eq}$ is the expected proportion of tracer particles moving with each passing surge and $\mu$ is as defined previously. For this class of probability models, the total surface concentration of tracer particles is independent of the specific probability laws governing erosional and depositional increment thicknesses. At equilibrium in this special situation, James ties in the average distance of tracer particle movement $E(\Delta X)$ to obtain the average longshore tracer velocity as $v_x = Pm_{eq}E(\Delta X)/\Delta t$, and the discharge Q as $\mu E(\Delta X)/\Delta t$. From this it emerges that $\mu$ is explicitly $Pm_{eq}/f_{eq}(0)$. But, more important, the final result, that $Q = v_x/f_{eq}(0)$, shows that the volume drift rate "is simply the tracer centroid velocity divided by the total surface concentration of tracer material, both of which are measurable quantities."

This example is presented in a very condensed manner here, but all the equations are included in James' paper. The final result in this example is step 6 in Table 4.1, and it illustrates that the specific question originally faced - how to estimate volume littoral drift rate in the field - becomes a special situation on a conceptual model that has other equally interesting theoretical implications.

## DISCUSSION AND CONCLUDING REMARKS

Perhaps the most important points about Table 4.1 are its implications regarding the design of quantitative geologic studies. Emphasis here is on the classic fields of geology rather than on strictly geophysical or geochemical research, where a body of analytical practice is relatively highly developed.

Methodology in quantitative geology has shown remarkable growth in the past several decades, especially in the increasingly rigorous use of conventional statistical techniques. This aspect itself received a major thrust as the high-speed digital computer became increasingly available during the early and middle 1960s. There is no question about the contributions made by the computer in removing the burdens of simple statistical computations and in opening paths for handling more than two or three variates simultaneously, almost completely unfeasible in precomputer days. One need only scan Burma's 1949 paper on discriminant analysis by hand methods to see how relatively hopeless the widespread adoption of multivariate methods was less than 30 years ago. Yet the seeds had been planted, and geologists with a statistical background were quick to adopt computer techniques.

As with every innovation, however, widespread availability of computer programs for multivariate analysis brought some potential dangers as well. Thus, it was no longer necessary to understand the underlying theory of discriminants, trend surfaces, stepwise regression, factor analysis, etc., in using the computer. Without that sense of judgment that adapts the method to the problem (rather than the other way about), the geologist may have his data "cooked" by a recipe not familiar to him. A danger here is that emphasis may be shifted unintentionally toward the realm of collecting data in search of a model, rather than for testing a formalized conceptual model that seeks to describe or explain a particular phenomenon.

This statement does not imply that standard least-squares models play no role in the scheme of sequential modeling given in Table 4.1. On the contrary, their role becomes more critical than in current usage. A conceptual model, in its first rough stage, requires some substantive knowledge about the nature of the phenomenon being studied. Preliminary field studies, accompanied by judicious sampling and measurement of variates intuitively selected, are normally used to acquire a "feel" for the phenomenon by noting how these variates are correlated with one another, whether they display systematic areal patterns, and whether they can be "sorted out" in some fashion that may disclose their relative value to the larger study.

In the franework of sequential modeling, most standard least-squares techniques are used as search models rather than as end products. An important exception is the predictor model, usually a regression equation, in which the immediate objective is to determine an optimum set of variates useful in estimating the numerical value of some other variate not directly measurable. A good example is provided in oil exploration, where observable combinations of lithologies, their thickness and sequence, their structural attitude, etc., are used to evaluate the oil potential of as yet unexplored areas or basins.

In some less directly applied aspects of geology, the predictor model is useful also as an end product. In many sedimentologic studies of Recent sediments, the objective is to discern criteria that help identify or interpret ancient sedimentary rocks. Widespread interest in distinguishing beach, river, and dune sands in the stratigraphic section is a good example. Here the process elements have vanished, and the observational data are confined to response elements locked into the ancient rocks. Classification procedures also represent an important application of the general linear model.

In terms of the three types of sands mentioned, it seems possible that a shift to sequential modeling may help to reduce seeming overlap of attributes and noise content in some observations. Lorenz's study does this in paleoecology, where all the observable data are response elements, but by judicious selection of inferred processes, a model is derived that opens the way for the types of critical observations that validly test the postulates. James' example is an illustration of the need for integrating process and response in the study of present-day processes, and it may well be that if sequential modeling is applied to the three types of sands as they are now being formed, criteria might emerge that would be more discriminatory than dependence on response elements only.

A final topic that deserves discussion is the choice between deterministic and probabilistic modeling in the scheme of Table 4.1. I mentioned that the sequence may be useful for both types of modeling, but that I prefer the probabilistic version partly as a matter of convenience. There is current confusion among geologists as to which is the "right" approach. The correct answer to this is probably neither and both; so much depends on the objectives of the study. The situation is further complicated by the unfortunate connotation the word "random" has for many physical scientists.

There are two aspects of randomness in geology that need to be kept separate in discussions of this sort. One has to do with the question of an element of "inherent uncertainty" in natural processes - a philosophy in which, as we penetrate natural phenomena more deeply, we reach a point where cause and effect lose all its meaning. The second view of randomness, with which I am concerned here, is willing to agree that natural phenomena are basically deterministic, but it also asserts that for complex phenomena, deterministic modeling is not always the optimum approach.

Dice rolling is not a geologic problem, but it can be used to draw a parallel. I do not know of any deterministic model that can exactly predict whether a 7 or non-7 will come faceup with a pair of dice; yet there must be some complex chain of cause-and-effect couplets that control the result. Gamblers have learned (with the help, incidentally, of some outstanding mathematicians) that a probabilistic approach is more satisfactory than investing money in supporting research on the deterministic elements that control the outcome of a toss. Probability theory based on the sample space of all possible outcomes yields a completely rigorous result that states that the probability of a 7 coming faceup is exactly 6/36.

Everyone is willing to accept the probabilistic approach in dice rolling, but many geologists seem unwilling to accept the same principle in explaining such a geologic phenomena as stream bifurcation in a growing stream network. One reason for this is that no geologist has taken the time to look at the total sample space for stream bifurcations, although it could certainly be approached by carefully designed field studies. What would emerge, in my opinion, is that bifurcation can occur as a result of numerous cause-and-effect couplets, but that after the event (especially if some years have elapsed), there may be no way of saying just which chain of events produced a particular bifurcation.

My argument simply is that many phenomena in geology are in this category where uncertainty is present regarding the exact set of circumstances that produce the given end product. If several series of events can result in identically the same end product, we lose that unique one-to-one functional relation between a specific control and a specific response that is characteristic of a deterministic model. In short, we lose the ability to make predictions about a dependent value except in a group sense, because we are now in the realm of sample space, which includes all combinations of the independent variates that could give rise to the same resulting value. Although we lose the ability to make predictions of a single event with probability $p = 1$, we gain some insight into partial dependencies that may be intertwined with the basic deterministic controls that drive the phenomenon as a whole.

Markov chains, independent-events models, random walks, and other conventional stochastic mechanisms play their role in this type of modeling by disclosing some aspects of the probabilistic drives that are or may be present in the process. To me, it is reasonable to assume that probabilistic elements are present, in part because natural phenomena display fluctuations that are larger than can be attributable simply to measurement error. Thus, I find it more convenient to start with the probabilistic mechanism that may be introducing the fluctuations, before committing myself to some specific deterministic core in the process.

## REFERENCES


Bharucha-Reid, A. T., 1960, Elements of the theory of stochastic processes and their applications: McGraw-Hill Book Co., New York, 468 p.

Briggs, L. I., and Pollack, H. N., 1967, Digital model of evaporite sedimentation: Science, v. 155, no. 3761, p. 453-456.

Burma, B. H., 1949, Studies in quantitative paleontology. II. Multivariate analysis - a new analytical tool for paleontology and geology: Jour. Paleontology, v. 23, no. 1, p. 95-103.

Coleman, J. S., 1964, Introduction to mathematical sociology: Free Press, Glencoe, Illinois, 554 p.

Howard, A. D., 1971, Simulation of stream networks by headward growth: Geog. Anal., v. 3, p. 29-50.

James, W. R., 1967, Nonlinear models for trend analysis in geology: Kansas Geol. Survey Computer Contr. 12, p. 26-30.

James, W. R., 1970, A class of probability models for littoral drift: Proc. 12th Intern. Coastal Eng. Conf., p. 831-837.

James, W. R., and Krumbein, W. C., 1969, Frequency distributions of stream link lengths: Jour. Geology, v. 77, no. 5, p. 544-565.

Krumbein, W. C., 1970, Geological models in transition in geostatistics, _in_ Geostatistics: Plenum Press, New York, p. 143-161.

Krumbein, W. C., and Graybill, F. A., 1965, An introduction to statistical models in geology: McGraw-Hill Book Co., New York, 475 p.

Krumbein, W. C., and LaMonica, G. B., 1966, Classification and organization of quantitative data in geology: Geologica Romanà, v. 5, p. 339-354.

Krumbein, W. C., and Dacey, M. F., 1969, Markov chains and embedded Markov chains in geology: Jour. Math. Geology, v. 1, no. 1, p. 79-96.

Leopold, L. B., and Langbein, W. B., 1962, The concept of entropy in landscape evolution: U.S. Geological Survey Prof. Paper 500-A, p. A-1-A-20.

Lorenz, D. M., 1973, Edenian (Upper Ordovician) benthic community ecology in north-central Kentucky: unpubl. doctoral dissertation, Northwestern Univ., 318 p.

McEwen, G. F., 1950, A statistical model of instantaneous point and disk cources with applications to oceanographic observations: Trans. Am. Geophys. Union, v. 31, no. 1, p. 33-46.

Schwarzacher, W., 1967, Some experiments to simulate the Pennsylvanian rock sequence of Kansas: Kansas Geol. Survey Computer Contr. 18, p. 5-14.

Shreve, R. L., 1966, Statistical law of stream numbers: Jour. Geology, v. 74, no. 1, p. 17-37.

Shreve, R. L., 1967, Infinite topologically random channel networks: Jour. Geology, v. 75, no. 2, p. 178-186.

Shreve, R. L., 1969, Stream lengths and basin areas in topologically random channel networks: Jour. Geology, v. 77, no. 4, p. 397-414.

Sievert, R. M., 1930, Die V-strahlungsintensität an der Oberfläche und in der nächsten Umgebung von Radiumnadeln: Acta Radiologica, v. 11, p. 239-301.

Smart, J. S., 1968, Statistical properties of stream lengths: Water Resources Res., v. 4, no. 5, p. 1001-1021.

Smart, J. S., 1969, Topological properties of channel networks: Geol. Soc. America Bull., v. 80, no. 9, p. 1757-1774.

Smart, J. S., 1971, Channel networks: Technical Rep. No. 4, ONR Contract N00014-70-C-0188, Task NR 389-155, 69 p.

Watson, R. A., 1969, Explanation and prediction in geology: Jour. Geology, v. 77, no. 4, p. 488-494.

# A Random-Walk Simulation Model of Alluvial-Fan Deposition

# 5

W. E. Price, Jr.

ABSTRACT

A digital model based on a random walk is used in an experiment to determine how well such a model is able to simulate alluvial-fan deposition. The model is three dimensional, and is dynamic with respect to both time and space. Two principal stochastic events are employed: (1) a relative uplift of the mountain area that is the source of the fan sediments, and (2) a storm event of sufficient magnitude to result in the deposition of material on the fan. These two events are assumed to follow independent Poisson processes with exponentially distributed interoccurrence times. The pattern of deposition is determined by a random walk from the canyon mouth at the mountain front, and each depositional event is assumed to occur instantaneously. The direction that each step in the walk takes is determined probabilistically by the gradient in the direction of flow, the momentum of flow, and the boundary conditions stipulated in the model. The type of flow - whether a debris flow, water flow, or eroding water flow - depends upon the thickness of erodible material in the basin. Deposition is assumed to occur over the entire length of the channel as a bad tapered in the direction of flow. Grain size of the water-flow deposits is governed by slope and velocity. Erosion is considered negative deposition and results from the exponential decline in elevation of the main stream channel at the fan apex or from water flows containing little basin sediment. Results of the simulation are printed as geologic and topographic maps of the fan surface and as geologic sections through the deposits.

## INTRODUCTION

The purpose of this paper is to describe the design and results achieved by using a digital model based on a random walk in an experiment to simulate alluvial-fan deposition. The model is discrete and three dimensional with respect to space and dynamic with respect to time. A plan view of the model grid system, and the boundary conditions assumed for some of the model runs, are shown in Figure 5.1. In the top center of the grid is the canyon mouth from which the flow debouches upon the basin

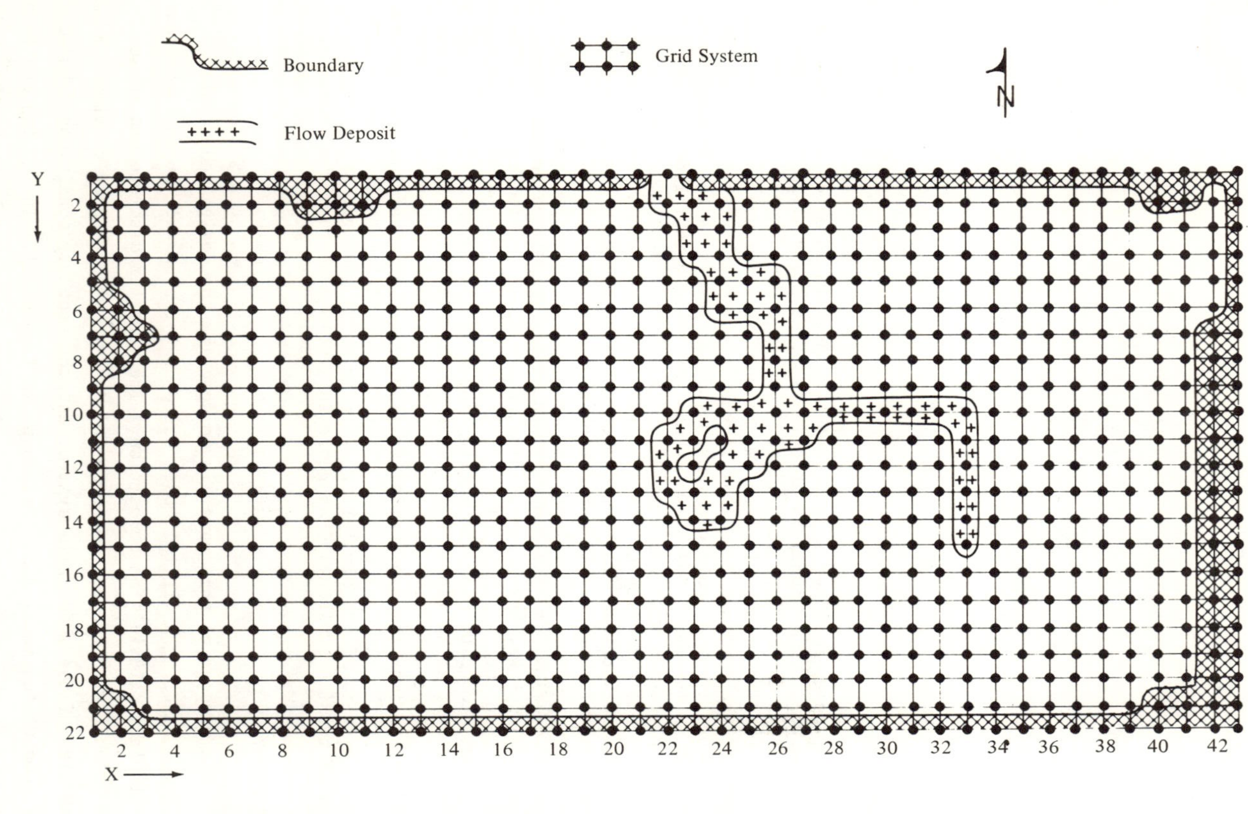


FIGURE 5.1 Map showing grid system, boundaries, and flow deposit.

of deposition in which the fan is built. The boundary along the top of the grid is the mountain front, which is bordered by the fault along which uplift of the mountain block occurs. On the sides of the grid are boundaries representing other fans or mountain areas. The boundary along the southern edge of the grid may be either a playa or a through-flowing stream.

## THE MODEL

The principal dynamic elements of the model are

1. Relative uplift of the mountain area containing the fan source basin
2. A storm that produces flow on the fan
3. Change in thickness of weathered material in the source basin
4. A random walk of storm flow into the fan
5. Deposition of sediment on the fan
6. Erosion both in the source basin and on the fan.

The first two of these dynamic events - relative uplift of the mountain area and storms that produce flow on the fan - are regarded as independent stochastic events. Relative uplift of the mountainous area is modeled by exponential functions that describe both the time of uplift and the magnitude of the earthquake presumed to have caused the uplift. Time of uplift is assumed to follow a Poisson process with exponentially distributed interoccurrence intervals of the form

$$f(t_u;\ \lambda_u) = \lambda_u e^{-\lambda_u t_u} \tag{5.1}$$

where $t_u$ = a period of time

$\lambda_u$ = the mean rate of occurrence of uplifts

Magnitude of the earthquake (Lomnitz, 1966) also is modeled by a negative exponential distribution of the form

$$f(M_e;\beta) = \beta e^{-\beta(M_e - M_0)} \tag{5.2}$$

where $\beta$ = a parameter related to b in the well-known formula of Gutenberg and Richter (1954, p. 17)

$M_0$ = minimum magnitude of earthquake events

$M_e$ = magnitude of a particular earthquake event

Storms by exponential distributions expressing the time and magnitude of the flow. Time of flow is given by

$$f(t_f;\lambda_f) = \lambda_f e^{-\lambda_f t_f} \tag{5.3}$$

where $t_f$ = a time period

$\lambda_f$ = mean rate of occurrence of flow events

Magnitude of flow (Shane and Lynn, 1964, p. 9-12) is given by

$$f(y;\gamma) = \frac{1}{\gamma} e^{-y/\gamma} \qquad \gamma > 0 \tag{5.4}$$

where $\gamma$ = mean peak flow rate and

$y$ = peak flow rate magnitude

The amount and type of material deposited on the fan during each flow event, however, are assumed to be independent of the magnitude of the flow, but dependent upon the volume of weathered material that is immediately available for erosion from the source basin. The areal extent of this material is considered constant throughout a model run, but its thickness increases with time according to the relation

$$y_s = m_s(1 - e^{-\eta t}) \tag{5.5}$$

where $y_s$ = thickness of the weathered layer

$m_s$ = a maximum thickness of the weathered layer

$\eta$ = a constant related to the rate of development of the weathered layer and

$t$ = a time period

A graph of this function is shown in Figure 5.2. Here, the thickness of the weathered layer is the ordinate and time is the abscissa. The curve shows the increase of the thickness of the weathered layer with time. Other aspects of the graph will be explained later.

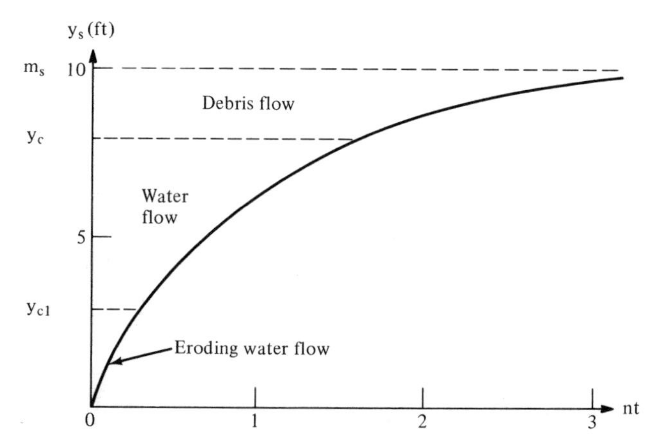


FIGURE 5.2 Graph showing rate of increase of weathered layer in basin and critical values for flow events in model.

Storms occurring over the model source basin produce runoff that removes the readily erodible material and brings it down to the canyon mouth where the random walk on the model grid system begins. Walks may proceed in any of four directions, but the probability of movement in any given direction ($P_d$) is proportional to the gradient (s) in that direction and is expressed by

$$P_d = 0.25 - 0.75s \tag{5.6}$$

Each of the transitional probabilities computed by this equation also is weighted by a momentum coefficient that expresses the fact that once a stream of water is headed in a given direction, it will tend to continue in that same direction. Figure 5.3 shows the method used in computing the transitional probabilities. In the north direction, a boundary is encountered. In the model, this is equivalent to a highly positive gradient, and therefore $P_N = 0.0$. In the east direction, the gradient is negative, and by Equation 5.6 we compute $P_E = 0.27$. Similarly, $P_S$ is computed as 0.29. To the west, a positive gradient again exists, and $P_W = 0.0$. Assuming that the previous direction of flow was to the east, and weighting the probabilities by a factor of 1.5, we obtain a final probability of 0.58 for movement to the east and 0.42 for movement to the south.

Two types of deposits were simulated: debris flows and water flows. Debris flows consist of poorly sorted material ranging in size from silt and clay to boulders, and are assumed to occur when the weathered layer in the source basin has attained a certain arbitrary thickness ($y_c$) (see Fig. 5.2) and a storm strikes the basin. If the weathered layer is less

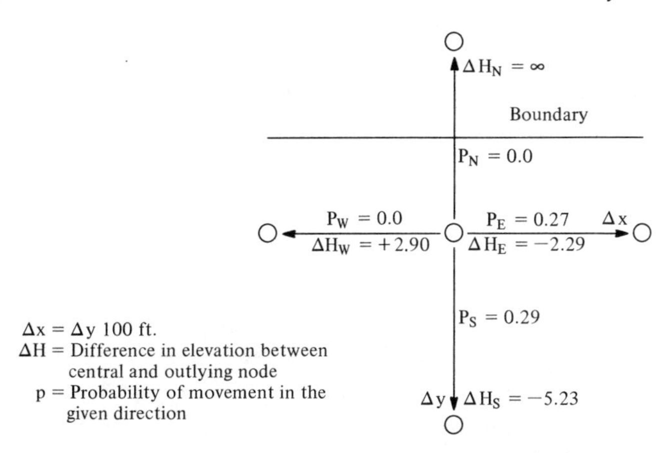


FIGURE 5.3 Diagram illustrating method of computing transitional probabilities.

than this thickness and a storm occurs, water flows will deposit moderately to well-sorted silty to gravelly sand, or will erode if the weathered thickness is equal to or less than $y_{c_1}$. The grain size of water-flow deposits is assumed to be directly related to slope and is expressed by

$$d = c_f s \tag{5.7}$$

where $d$ = median grain size
$c_f$ = a coefficient
$s$ = slope

Deposition is assumed to occur instantaneously in time over the entire length of the channel. Deposits may develop irregular shapes or may branch, and their thickness tapers in the direction of flow.

Two types of erosion are modeled. The first type, expressed by

$$h = h_0 e^{-kt} \tag{5.8}$$

where $h$ = the elevation (in feet) of the rock stream channel above base level at a particular time $t$
$h_0$ = the elevation of the rock stream channel immediately following an uplift at $t_0$
$k$ = a dimensionless parameter expressing the rate of decline of the stream channel

results from the downward cutting of the main stream into its rock channel above the fan. The second type, erosion of sediments from the fan itself, is due to water flows containing little basin sediment, and is treated as negative deposition.

Some examples of simulated flow events and fans are shown in Figures 5.4 to 5.6. Figure 5.4 illustrates the pattern formed by a few debris and water flows on a flat surface. A topographic map of the surface of a fan simulated by 32 random walks, each of which is a water flow, is shown in Figure 5.5. Particle-size graduation of sediments in the same fan is shown in Figure 5.6.

## RESULTS

Results of this study indicate that a digital model based on a random walk can be used to simulate alluvial-fan deposition. The general form of the deposits is that of an alluvial fan, the pattern of simulated flows resembles that of real flows, and fan facies show a concentration of debris flows near the apex and a particle-size decrease of water-flow deposits downfan. The model holds promise for future development, and future

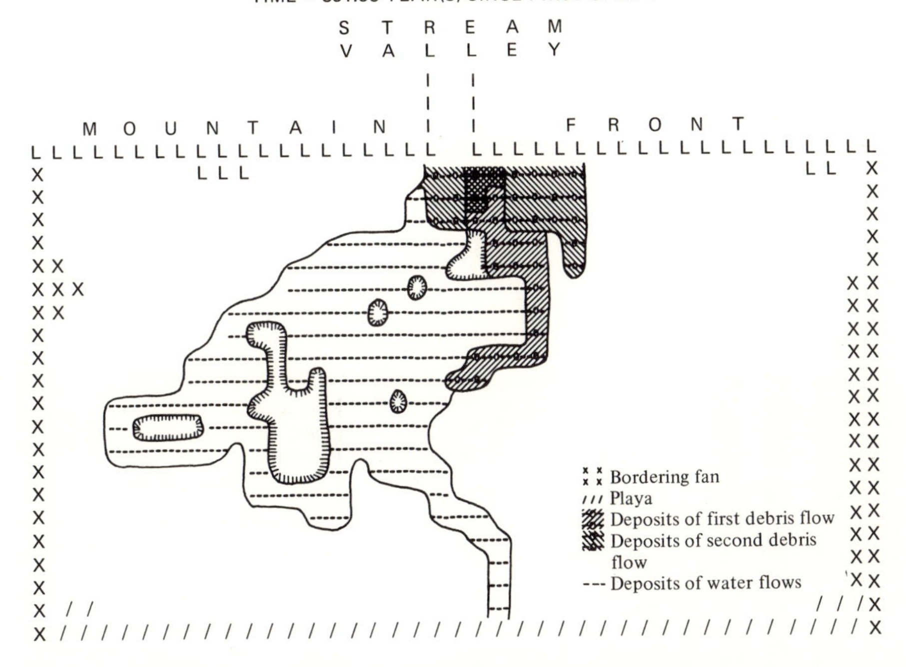


FIGURE 5.4 Computer printout showing shape of simulated debris-flow and water-flow deposits.

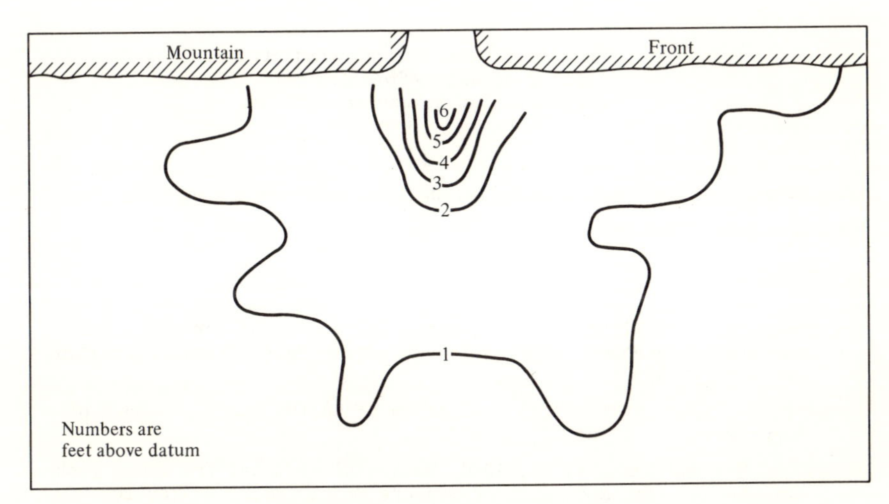


FIGURE 5.5 Topographic map of simulated fan consisting of water-flow deposits.

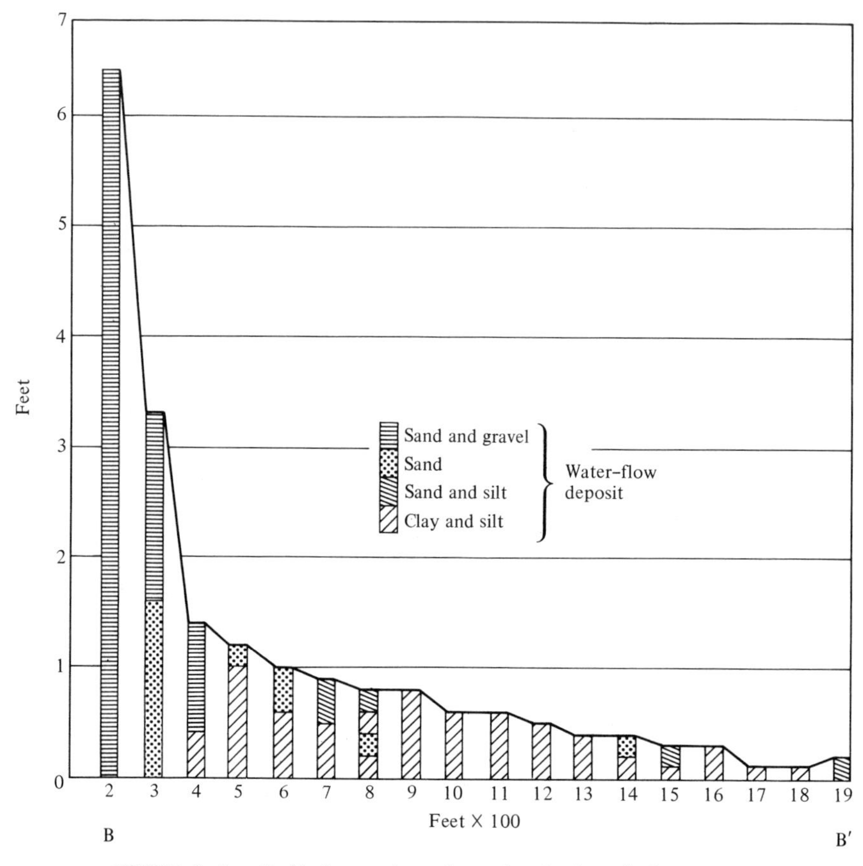


FIGURE 5.6 Radial section through simulated fan consisting of water-flow deposits. Section perpendicular to mountain front.

work should include quantitative validation, sensitivity analysis, and further experimentation.

## REFERENCES


Gutenberg, B., and Richter, C. F., 1954, Seismicity of the earth and associated phenomena (2nd ed.): Princeton Univ. Press, Princeton, New Jersey, 310 p.

Lomnitz, C., 1966, Statistical prediction of earthquakes: Reviews of Geophysics, v. 4, no. 3, p. 377-393.

Shane, R. M., and Lynn, W. R., 1964, Mathematical model for flood risk evaluation: Am. Soc. Civil Engineers Proc., Jour. Hydraulics Div., v. 90, no. HY6, p. 1-20.

# Sedimentary Porous Materials as a Realization of a Stochastic Process

F. W. Preston and J. C. Davis

ABSTRACT

Statistical variation in the properties of porous materials, including sedimentary rocks, has been recognized for over half a century. In early studies, determinations were made of measures of central tendency and variation in the sizes of pores and grains. Some work was directed to measuring probability density functions of pore- or grain-size distributions. The effort needed for accurate determination of such functions precluded their widespread use. Despite the early recognition of the statistical nature of porous materials, fluid-flow models of porous media continue to be based on simple deterministic models of spherical particle or capillary tube assemblages. Such models are inadequate for the study of fine details in pore-grain structure or fluid flow at the particulate level. These models are equally unsuited for studies of particle transport and sedimentation or for the characterization of sediment and sedimentary rocks. The models provide no basis for accurate lithologic description or analysis.

Recently, experimental and theoretical tools have been developed for the description of porous materials as a realization of a stochastic or random process. Methods of Fourier optics and optical data processing, both digital and analog, combined with the theory of random processes, constitute the basis for a new approach to analyzing and simulating mathematically porous materials.

Representation of porous material as a realization of a random process implies that each sample in a population will be different from others in a point-to-point sense, yet be similar in average properties. If the process that generates a given sample (a particular realization) is the same for three coordinate directions, the material is considered isotropic and homogeneous. The concept of statistical homogeneity thus is related to the equivalence of the random process for all coordinate directions. Similarly, it is possible to describe precisely trends in statistical parameters through long or short intervals by unequal variation in these random processes for the three coordinate directions. As a demonstration of the sufficiency of the parameters extracted from a porous material, it is possible to simulate mathematically a structure with all the attributes of a natural or synthetic porous material.

Fourier transforms of images of the pore-grain structure are useful for classifying pore structure numerically and for determining the pore- or grain-size distribution in a sample quickly, cheaply, and with precision. These same transforms permit, through optical filtering, the analysis and synthesis of "idealized" porous materials. Such analysis or synthesis can be done with optical image analysis equipment in either digital or analog form. The necessary equipment for optical data processing as a method of pore-structure or grain-size analysis is relatively inexpensive, although radically different from traditional measurement devices.

## THE NATURE OF SEDIMENTARY BEDS

Geologists group sedimentary rocks within the stratigraphic succession into progressively smaller units of increasingly greater homogeneity. A formation is perhaps the most widely used of the larger stratigraphic subdivisions, defined as a body of sedimentary rock that forms a mappable lithologic unit. This definition implies that a formation consists of rock of sufficiently unique character to allow it to be distinguished from overlying and underlying formations, and that it is sufficiently uniform to be recognizable throughout its extent.

However, the degree of internal uniformity is only relative, and a formation may contain finer subdivisions of widely differing lithologies. These subdivisions are referred to as "beds." Many definitions of a sedimentary bed have been given, but it is generally agreed that they consist of tabular bodies of uniform composition, separated from enclosing beds by more or less distinct boundaries. Describing a unit as a bed implies that it was formed throughout its extent during a single depositional episode of essentially constant physical and chemical conditions. Boundaries between beds reflect changes in the depositional regime. A succession may consist of identical beds, distinguishable only by the presence of discontinuities between them. In other successions, beds of different lithologies will alternate, forming sequences of great complexity. Note that bed and formation are based on different criteria, the former on internal consistency and the latter on mappability. A formation may consist of a single bed or a large collection of beds.

Within a single bed, the rock may be entirely homogeneous and seemingly isotropic, in which situation it is referred to as massive. Heterogeneity may be expressed as a graduation in lithology from bottom to top of a bed, or as a series of minor alterations in composition or arrangement of constituents within the bed. The latter variations are expressed as laminations and generally are regarded as reflecting minor fluctuations within the depositional regime.

The assumption that a single bed results from consistent environmental conditions is an important one, corresponding to Otto's (1938) definition of a sedimentation unit. Because depositional conditions presumably were stationary or gradually evolutionary with no abrupt alterations that would create a bed boundary, the physical features of a bed that reflect environmental processes should be uniform or have a regular trend. Variance in physical characteristics within a bed can be attributed to random fluctuations in the complex of depositional controls.

By regarding a bed as a single population with a true mean and variance, we can extrapolate from small samples to the bed itself, and may draw inferences about the environment in which the bed was created. The various designs for sampling sedimentary units, such as those described by Griffiths (1967), have as their objective the elimination of bias in sample collection. They are designed to detect possible trends within a bed or increases in variance caused by inadvertently sampling from more than one population, such as would result from collecting across a bedding plane between two units.

However, this theoretical model may be inadequate, as was pointed out by Matheron several years ago (1962). We know from experience that measurements made on sediments are not independent of sampling locations within a bed, nor are they independent of the size, shape, and orientation of the samples. Any procedure that regards a sedimentary bed as a homogeneous universe from which independent samples may be plucked at random may yield misleading results if the population is not, in fact, homogeneous and isotropic.

The model assumed for a sedimentary bed determines the nature of the variables that can be measured. Griffiths (1952, 1961), for example, suggests that a sedimentary rock can be regarded as a function of five fundamental properties: the mineralogy, size, shape, orientation, and packing of constituent grains.

In principle, pore geometry or a derived function, such as porosity or permeability, can be described by specifying interrelationships among these parameters for all grains in a sample of rock. However, there are conceptual problems in the definition of some of these parameters, as well as immense practical difficulties in their measurement. For example, except by volumetric determination, it is impossible to specify the size of a grain without defining its shape. Orientation also must be defined with reference to grain shape, as it involves the angles between grain axes and external reference directions. Both size and shape are properties of individual grains. Orientation is a property of a grain, but only within the context of an aggregate. Packing is a property that has meaning only for an aggregate, as an individual grain has no "packing." Indeed, it seems difficult to define packing in any unambiguous manner (cf., Emery and Griffiths, 1954; Kahn, 1956a, 1956b), although this is the parameter that pertains most directly to the geometry of voids within a sandstone or other porous aggregate. Although size and shape can be measured for individual grains by resorting to operational definitions of measurement procedures, orientation and packing can be assessed only by considering a volume of consolidated rock.

Difficulties inherent in the definition and measurement of conventional parameters of size, shape, orientation, and packing have led us to consider the utility of vectorial relationships for the description of rock properties. Within this conceptual framework, rock measurements are regarded as spatially related, the degree of relationship being a function of the distance between samples and the orientation of the sampling pattern. The directional dependence of measured properties may be defined as an expression of their vectorial character. Some properties, such as permeability, have always been regarded as vectorial in nature. However, it is possible to extend this concept to all types of measurements made on sediments, including those traditionally gathered in a petrographic study. By adopting this concept, certain difficulties inherent in conventional approaches may be avoided, and measurements more characteristic of the rock as a whole can be made.

In a conventional statistical analysis of a property measured on a sedimentary rock, it is sufficient to estimate the probability density function of the property over the entire bed. A vectorial approach, however, also requires the estimation of the joint probability distribution function for all possible pairs of points throughout the bed. If the rock is anisotropic in the property considered, these probability functions will change with the vectorial direction. If the rock is heterogeneous, both the probability density functions and joint probability density functions change with distance along any given vector.

## VECTORIAL PROPERTIES OF SANDSTONE

A property T can be defined at any point i along an arbitrary sampling line passed through a rock in a vector direction $\vec{h}$. As an example, consider Figure 6.1, which represents a line drawn across a thin section of a sandstone. Every point on the line is located within either a pore

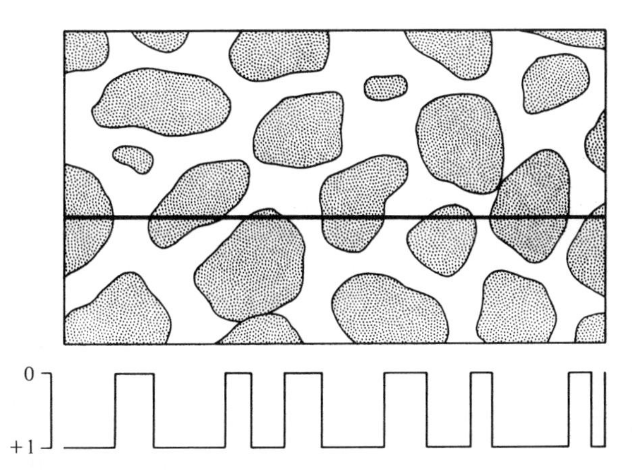


FIGURE 6.1 Idealized image of sandstone thin section. Stippled areas represent grains. Traverse, shown as heavy line, will create square-wave function shown at bottom, where grain is represented as +1, pore as 0. (After Preston and Davis, in press.)

or a grain. If a grain is denoted by +1 and a pore by 0, the trace of the line forms a square wave as it alternately passes from grain to pore and back again. (A similar function was previously defined by Fara and Scheidegger in 1961.) The nature of this sampling function, which we shall call Y, reflects the size distribution of grains, their orientation and their packing, as well as the nature of the voids within the sandstone. No tractable deterministic expression could describe the complex pattern of measurements that would emerge from all possible lines drawn at all possible orientations across a thin section. However, the function Y can be described in a statistical sense.

In practice, Y is determined by preparing a high-contrast photographic image of a thin section whose pores have been filled with red epoxy (Minoura and Conley, 1971). The image is scanned by a microdensitometer that records the optical density of the film at sequential points along successive lines. The resulting values are clipped to yield a discrete record of pore versus grain.

With this or a similar sampling technique, a function T may be defined as a random variable existing at equally spaced points, $i = 1, 2, \ldots, n$, along a line. Let the separation between successive points be $\Delta x$. At each point i, the variable $T_i$ can assume a range of values determined by the probability density function $f_i(T)$, giving the permissible distribution of T at point i. In the particular example discussed in this paper, the function can assume only the values $Y = \{^1_0$. The ordered set $\{T_i\}$, $i = 1, 2, \ldots, \infty$, is called a stochastic process. The most general definition of a stochastic process allows different probability density functions at each point i, and requires joint probability density functions $f_{i,j}(T_i,T_j)$ for any arbitrary set of points $i,j = 1, 2, \ldots, \infty$. The function T is doubly infinite as there exists an infinite number of points along the line, and at each point i there is an infinite set of values that $T_i$ can assume. This doubly infinite set of functions is called an "ensemble." The ensemble thus defined is general but estimation of the probability density function and joint probability density functions becomes extremely laborious.

If it may be assumed that the probability density function $f_i(T)$ is the same for all i, the stochastic process T is statistically stable and is said to be stationary. Stationarity implies that the joint probability density functions for the various $T_i$'s become functions only of the distance between points rather than functions of i. Thus, in a realization $\{T_i\}$ from a stationary stochastic process, we can consider the n-k possible pairs of values $(T_1,T_{i+k})$, $i = 1, 2, \ldots, n-k$. These constitute observations from the joint probability density function $f(T_i,T_j)$ where the points i,j are a distance $k\Delta x$ apart.

Although stationarity is frequently assumed, it is possible to have nonstationarity in either the mean or variance, or both. Nonstationarity in the mean may arise from the presence of periodicities or trends. Both stochastic and deterministic trends may occur. A sedimentary rock unit may be considered to be stochastically stationary if the pattern of variation in a property is similar in each sampled subarea of a bed.

For stationary stochastic processes, an important property is the autocovariance function ACVF(k). For equally spaced data $T_i$, $i = 1, 2, \ldots, \infty$, this is defined as

$$\mathrm{ACVF}(k) = E\,[(T_i - \mu)\,(T_{i+k} - \mu)] \equiv \mathrm{COV}\,[T_i, T_{i+k}] \tag{6.1}$$

where E is expectation, COV is covariance, and $\mu$ is the population mean of $T_i$, assumed to be constant for the ensemble. The integer k is called the lag. Under most circumstances, ACVF(k) must be estimated from a finite data set without prior knowledge of $\mu$. The estimation equation is

$$\mathrm{ACVF}(k) = \frac{1}{n} \sum_{i=1}^{1=n-k} (T_i - \overline{T})\,(T_{i+k} - T) \tag{6.2}$$

$$\overline{T} = \frac{1}{n} \sum_{i=1}^{i=n} T_i \tag{6.3}$$

An algebraically equivalent but computationally more efficient form is

$$\mathrm{ACVF}(k) = \frac{1}{n}\left[\sum_{i=1}^{i=n-k} T_i T_{i+k} - \overline{T}\sum_{i=1}^{i=n-k}(T_i + T_{i+k}) + (n-k)(\overline{T})^2\right] \tag{6.4}$$

An autocovariance function may be converted to dimensionless form as the autocorrelation function ACF, where

$$\mathrm{ACF}(k) = \frac{\mathrm{ACVF}(k)}{\mathrm{ACVF}(0)} \tag{6.5}$$

The autocovariance function describes the degree to which points distance $k\Delta x$ apart are related and may be used to describe time series. However, physical interpretation of the ACVF may be difficult or impossible (Tukey, 1966; Jenkins and Watts, 1968). Usually it is simpler to interpret the power spectrum of the autocovariance function (Tukey, 1970). A widely accepted estimate of the power spectrum of a stochastic process may be obtained as the finite Fourier transform of the autocovariance function:

$$S_j^2 = 2\Delta x\left[\mathrm{ACVF}(0) + 2\sum_{k=1}^{k=m-1}\mathrm{ACVF}(k)\,W(k)\cos\frac{\pi k j}{F}\right] \tag{6.6}$$

where $\Delta x$ = distance between successive data points

$W(k)$ = smoothing function (lag window in the terminology of Blackman and Tukey, 1958 ) used to produce a more satisfactory continuous function $S_j^2$ from the discrete values

$k$ = lag index

$m$ = maximum lag (truncation point of Jenkins and Watts, 1968 ), an integer $m < (N/10)$

$F$ = an integer multiple of m, permitting interpolation of additional spectral values without additional ACVF values; generally, $F \lesssim 4m$

$j$ = frequency index; $j = 0, 1, 2, \ldots, F$

Changes in distance can be treated in a manner analogous to changes in time. Although Figure 6.1 represents a transverse across a thin section, it can be regarded as a "time" series in which periods are measured in millimeters and frequencies are measured in cycles per millimeter. The spatial frequency corresponding to a given index j in Equation 6.6 is

$$f_j = \frac{1}{2\Delta x}\,\frac{j}{F} \tag{6.7}$$

Determination of an appropriate lag window is a significant experimental problem. So much effort has been invested in this pursuit that it has been called window carpentry by Blackman and Tukey (1958). At least five different windows have been widely tested on a variety of data sets (Marks, 1969). However, there is little difference between them when applied to data series, such as $Y_i$, extracted from sandstones. All windows smooth the data poorly if few points are available and large values are chosen for truncation points.

The Parzen window has some advantages in that it precludes negative values of $S_j^2$, which may arise with other windows and certain data sets. The Parzen window is given by

$$W(k) = 1 - \frac{6k^2}{m^2}\left(1 - \frac{k}{m}\right) \qquad 0 \leqslant k \leqslant \frac{m}{2}$$
$$W(k) = 2\left(1 - \frac{k}{m}\right)^3 \qquad \frac{m}{2} \leqslant k \leqslant m \tag{6.8}$$

The discrete power spectrum $S_j^2$, obtained from digitized photographs of sandstones, provides a satisfactory descriptor set for numerical classification of porous sandstones (Preston and Davis, in press). In the terminology of pattern recognition (Kanal and Chandrasekaran, 1969), the set $\{S_j^2\}$ constitutes an effective set of image "features."

Figure 6.2 shows four selected images of thin sections prepared from sandstones that form petroleum reservoirs. These images are part of a larger suite of reservoir samples being investigated in a research project sponsored by the American Petroleum Institute. The four sandstones illustrated are quartz arenites; they differ in grain size and grain shape, especially degree of rounding. Images were digitized at a spacing of $\Delta$ = 25 microns, and power spectra computed for a series of parallel sampling lines across each image. Resulting spectra are shown in Figure 6.3. Although the spectra are shown as continuous curves, they actually consist of sets of 40 discrete values $\{S_j^2\}$.

The efficacy of the power spectrum as a descriptor of the function Y can be assessed at least rudimentarily by considering how well the spectra serve to distinguish different samples. The first 16 (low-frequency) spectral values were used as variables, as higher frequencies contribute little to the total variance in the sample suite. The samples, which consist of five replicates of seven different sandstones, were projected onto the principal axes of the correlation matrix (Figure 6.4). The first two axes explain 63 and 35% of the original variance in the samples, or 98% of the total variance. The replicates form relatively distinct, compact clusters. Note that samples D and G are both St. Peter Sandstone and

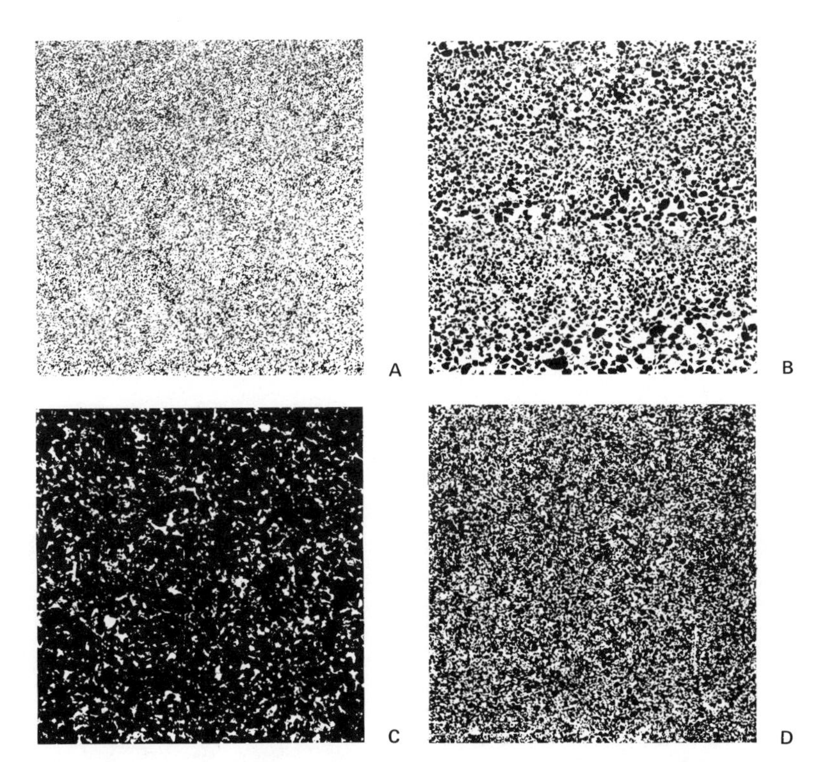


FIGURE 6.2 Photomicrographs of sandstones from units forming petroleum reservoirs. Sand grains appear black; pores are white. Magnification x 2. A. Woodbine Sandstone (Cretaceous), Texas. B. Woodbine Sandstone (Cretaceous), Texas. C. Gaskell Sandstone (Eocene), California. D. St. Peter Sandstone (Ordovician), Iowa. (After Preston and Davis, in press.)

closely overlap. Sandstones E, F, and G are not illustrated in Figure 6.2 but are visually similar to D. Sandstones A, B, and C are visually the most disparate of the sample suite.

Having found the ACVF, power spectrum, probability distribution, or other characteristic function of the variable T measured along oriented lines, we may regard the vector $\vec{h}$ as possessing an orientation with respect to an arbitrary axis system and to contain the characteristic function. In general, the function characterizing $\vec{h}$ will change with changes in orientation, although for an isotropic material the functions will be the same for all possible orientations of $\vec{h}$. The power spectrum of a stationary random process consists of a simple exponential decay curve. Spectra computed from porous materials have this same general characteristic, so it seems reasonable to assume that the sampling function $Y_i$ is the realization of a stationary random process. However, properties

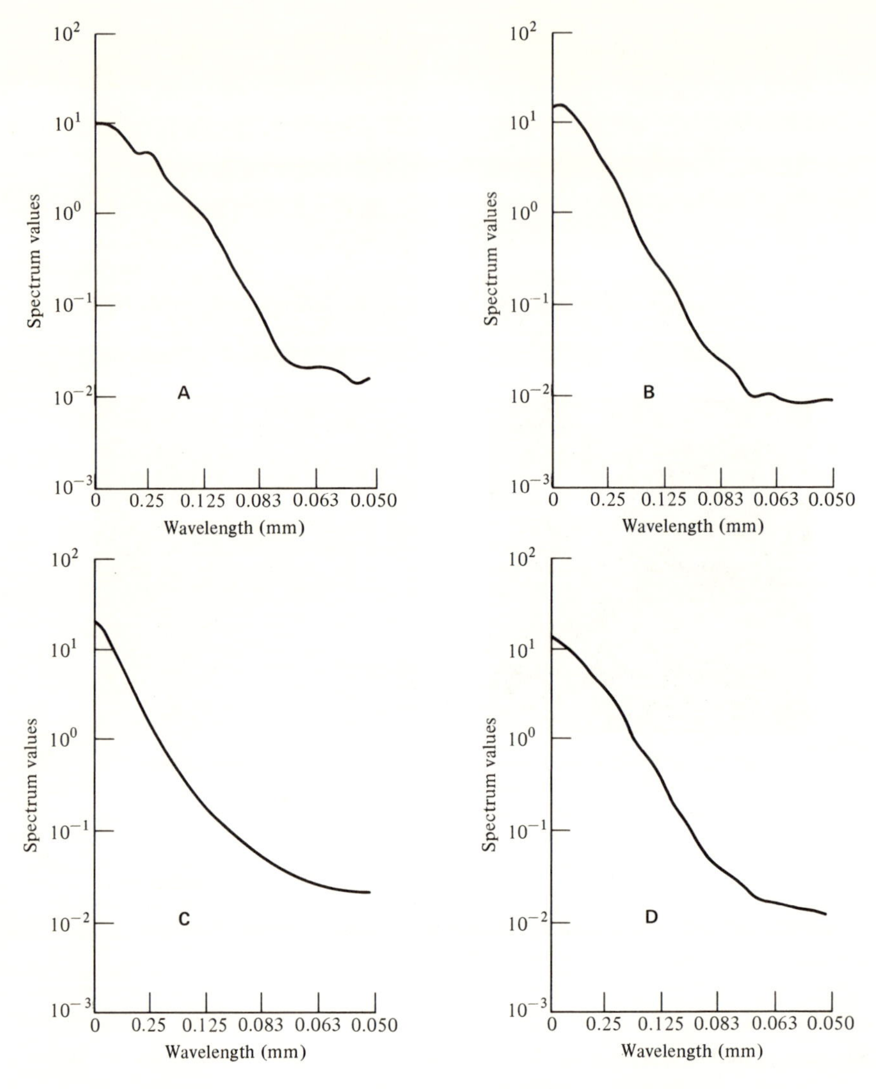


FIGURE 6.3 Power spectra of digitized traverses across images of sandstones; spectra correspond to images A, B, C, and D in Figure 6.2. (After Preston and Davis, in press.)

of the function $Y_i$ are, in general, different for different vector orientations, as a consequence of such factors as grain orientation and elongation of pores in certain directions. Therefore, the power spectrum derived from a sandstone may differ with orientation of the sampling vector. If the sample is anisotropic but homogeneous, the power spectra for all possible vector directions $\vec{h}$ will have the form of an ellipsoid, with higher values in the center and gradually decreasing power levels outward. A

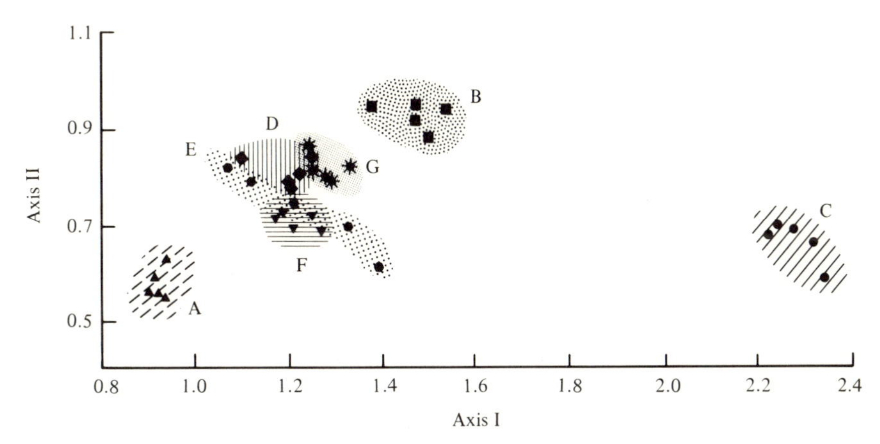


FIGURE 6.4 Projection of power spectra onto their first two principal axes. Five replicate spectra for each sandstone are shown by different symbols. Letters correspond to those in Figure 6.2. (After Preston and Davis, in press.)

heterogeneous sample also will yield an ellipsoidal three-dimensional power spectrum unless heterogeneity is expressed as uniformly spaced thin laminations or other highly periodic variations. The spectrum then will form a more complex solid.

It would be extremely useful if a series of one-dimensional power spectra taken along sample vectors could be resolved into the principal axes of the power spectrum ellipsoid. To show how this might be done, we will first consider a somewhat simpler problem.

Suppose we measure some property T along certain arbitrary vector directions $\vec{h}_i$. The property might be, for example, the maximum diameters of grains measured parallel to $\vec{h}_i$. For each vector direction $\vec{h}_i$, we can compute a mean maximum diameter. The vectors $\vec{h}_i$ then are characterized by their orientation with respect to an arbitrary system of reference axes (for convenience, this system might be $X_1$ = east-west, $X_2$ = north-south, and $X_3$ = up-down), and have magnitudes equal to the mean maximum grain diameter measured in the vector direction. A series of these vectors can be resolved into an ellipsoid whose principal axes correspond to the orientation of the principal axes of the grains and whose magnitudes reflect the relative lengths of the mean axes of the grains.

Denote as $\overline{T}_i$ the mean grain size measured in direction $\vec{h}_i$. Vector $\vec{h}_i$ forms the angle $\alpha_i$ with reference axis $X_1$, angle $\beta_i$ with reference axis $X_2$, and angle $\gamma_i$ with reference acis $X_3$. The length of each vector $\vec{h}_i$ is defined as equal to $\overline{T}_i$. We may fit an ellipsoid to the values of $\overline{T}_i$ by the equation

$$\overline{T}_i = C_{11}\cos^2\alpha_i + C_{22}\cos^2\beta_i + C_{33}\cos^2\gamma_i + 2C_{12}\cos\alpha_i\cos\beta_i + 2C_{13}\cos\alpha_i\cos\gamma_i + 2C_{23}\cos\beta_i\cos\gamma_i \quad (6.9)$$

which relates points on the surface to the principal axes of the ellipsoid. From a series of sample vectors $\vec{h}_i$, each characterized by a mean grain size $\overline{T}_i$, the principal axes of the ellipsoid describing mean grain size can be found by least squares. The linear equation can be premultiplied by its transpose to yield a 6 x 6 set of simultaneous equations A . The six unknown components $C_{ij}$ (i,j = 1, 2, 3) then can be factored. A vector of products between the $\overline{T}_i$ and the cosine terms forms the right-hand part of the equation; then

$$[A] \cdot [C] = [\overline{T}_i] \quad (6.10)$$

or

$$[C] = [A]^{-1} [\overline{T}_i] \quad (6.11)$$

We now know the coordinates of the axes of the best-fit ellipsoid with respect to the reference axes. These can be set in the form of a real symmetric 3 x 3 matrix of the form

$$\begin{bmatrix} C_{11} & C_{12} & C_{13} \\ C_{12} & C_{22} & C_{23} \\ C_{12} & C_{23} & C_{33} \end{bmatrix}$$

The matrix defines a series of three vectors in space whose lengths are proportional to the quantity $\overline{T}_i$ in that direction. The vectors constitute the principal axes of an ellipsoid; eigenvalues of the matrix define the lengths of these principal axes and represent estimates of the mean axes of grains in the sample space.

In practice, the property $\overline{T}$ may be any measurement with a vectorial component $\vec{h}$. This method has been used to determine the principal axes of the permeability ellipsoid of reservoir rock (Greenkorn and Johnson, 1964) and solid (Maasland, 1957; Maasland and Kirkham, 1959). In general, axes of ellipsoids reflecting different rock properties will not coincide, although we might logically expect a relationship between certain of them, such as moments of the grain-size distribution.

Rather than reducing the measurements along the vector $\vec{h}_i$ to a single value, such as $\overline{T}_i$, we may regard the vector as containing a function such as the power spectrum. That is, each vector $\vec{h}_i$ is characterized by a series of parameters $T_{ij}$ of the function in the vectorial direction. The problem is to resolve a collection of these parameters determined along a series of arbitrary vectors $\vec{h}_i$ into parameters along principal axes.

One possible approach is to fit an ellipsoid to only one feature of the functions, perhaps their first moment, and determine the principal axes from these. The assumption may be made that the principal axes of the first moment also are principal axes of all classes in the function. The function can be projected onto these principal axes using the weightings used to project the means.

Some insight into the nature of a three-dimensional power spectrum of a sandstone can be obtained by optical processing. A two-dimensional intensity spectrum, which is proportional to a continuous two-dimensional power spectrum, can be created from a thin-section image by using a simple optical system. The methodology has been described by many authors; one of the most succinct discussions is given by Shulman (1970). Pincus and his associates (see Pincus, 1969, for complete references) and Preston and Davis (in press) have examined the power spectrum of rock thin sections.

An optical processor uses a beam of coherent, collimated, monochromatic light to create a Fourier transform of an image. If such a beam is passed through a thin-section image, the grain-pore boundaries cause diffraction of the light. The angle of diffraction is a function of spatial dimensions of grains and pores in the image. Diffracted light may be collected by a lens and focused in the back focal plane; with appropriate optical geometry, intensity of light in this plane is related to the intensity of light emerging from the image by

$$I(x_0,y_0) = \frac{A^2}{\lambda^2 Z^2} \left| \iint t(x_1,y_1) \exp\left[-j \frac{2\pi}{\lambda Z} (x_0 x_1 + y_0 y_1)\right] dxdy \right|^2 \qquad (6.12)$$

where
$A$ = amplitude of incident light wave
$j$ = imaginary unit, $\sqrt{-1}$
$Z$ = distance between the image plane $x_1,y_1$ and the focal plane $x_0,y_0$
$x_0,y_0$ = coordinates of point 0 in transform plane
$x_1,y_1$ = coordinates of point 1 in image plane
$t(x_1,y_1)$ = amplitude of light falling upon point where coordinates are $x_1,y_1$
$\lambda$ = wavelength of light

This relationship is analogous to the square of a Fourier transformation, except for a constant phase factor. Intensity of light at any specific point in the focal plane represents the amount of a specific size detail in the input, or, in other words, the specific spatial frequency content of the image. If intensity of light in the focal plane is measured in a linear manner, the resulting intensity spectrum is directly proportional to the two-dimensional power spectrum of the input image.

The optical power spectrum of a random assemblage of apertures appears as a circular or elliptical field of light whose intensity decreases gradually outward from the center (Figure 6.5). This is the pattern usually seen in optical transforms of images of sandstone thin sections. Radial distance outward from the center of the transform is related to spatial wavelength in the image by

$$y_0 = \frac{\lambda f}{2\pi} \omega \tag{6.13}$$

where $\omega$ = spatial wavelength in image

$\lambda$ = wavelength of light

$f$ = focal length of lens system

$y_0$ = distance in the focal plane outward from transform center

FIGURE 6.5 Optical intensity spectrum as it appears in Fourier transform plane (back focal plane) of optical processor. Elongation of light pattern reflects different spatial frequency distributions along different directions in original image.

By this relationship, the optical intensity spectrum may be rescaled to correspond to power spectra computed by Fourier analysis.

If intensity is measured along an arbitrarily oriented radius across an optical intensity spectrum, the measurements are proportional to a one-dimensional power spectrum. This spectrum represents all possible parallel sampling lines across the original image that are oriented in a direction perpendicular to the arbitrary radius. Therefore, it is a physical analog of a vector $\vec{h}$ containing a power spectrum, as it is characterized by both orientation and a contained distribution. Figure 6.6 shows an optical intensity spectrum contoured in units that are proportional to log intensity. The arbitrarily oriented vectors $\vec{h}$ are represented by a series of fine lines, each of which contains a power-spectrum characteristic of the pore-grain property in a direction complementary to the vector. Common points, such as the 25% log intensity level, lie at the intersection of each vector $\vec{h}_i$ with the ellipse represented by the 25% contour line.

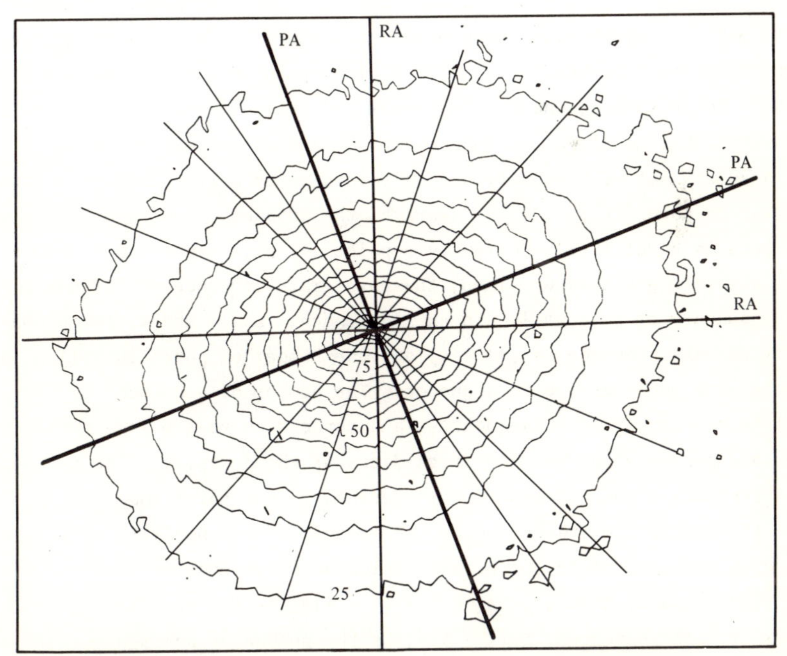


FIGURE 6.6 Power spectrum in Figure 6.5 contoured at equal intervals of percent log intensity. RA = reference axes of spectrum and image. PA = principal axes of two-dimensional spectrum, which are projections of the principal axes of power-spectrum ellipsoid. Light lines represent arbitrary vectors $\vec{h}$.

Relationships between principal axes of the ellipse, reference axes, and sample vectors can be readily seen in this physical realization.

Thin sections prepared along mutually perpendicular planes through a sandstone will provide a series of two-dimensional optical intensity spectra. From these, an ellipsoidal representation of the power spectrum of pore-grain relations in the sample can be determined. Principal axes of the ellipsoid may be determined directly, because each two-dimensional spectrum contains projections of the principal axes that can be located by inspection. If, however, the data consist of spectra computed along arbitrary vectors $\vec{h}_i$, orientation of the principal axes of the ellipsoid must be determined by least-squares fitting in the manner discussed previously.

## SYNTHESIS OF STOCHASTIC MODELS OF POROUS MATERIALS

If porous materials may be regarded as realizations of a stochastic process, it may be possible to create more realistic mathematical models of the pore structure than those based on assumed assemblages of spheres or on capillary tube bundles. An initial step in this direction has been taken by Roach (1968), who created pore- and grain-size distributions in one-dimensional models by an independent random stochastic process.

Assume a set of N equally spaced points along a line through a porous material of porosity $\phi$. On the average, the probability of a point lying in a pore is equal to $\phi$. Designate a point lying in a pore as P and one lying within a grain as G. Then, the pore-grain sequence PG defines a grain boundary or pore wall. In an independent random model, the probability of this sequence is $\phi(1 - \phi)$ and the number of pores is $N\phi(1 - \phi)$ because each pore has one PG sequence. The number of sequences GP is equal to the number of PG sequences, and hence we have the same number of pores and grains along any line regardless of the porosity. The sequence GPG represents a pore of length $1\Delta x$ where $\Delta x$ is the point spacing. Because the points are independently determined, the probability of this sequence is the product of the probabilities of occurrence of a grain, a pore, and a grain; that is, $\Pr(GPG) = (1 - \phi)\phi(1 - \phi)$, or $\phi(1 - \phi)^2$. The number of such sequences, $N\phi(1 - \phi)^2$, is the number of pores of length $1\Delta x$ along the line of N points. Similarly, there are $N\phi^2(1 - \phi)^2$ pores of length $2\Delta x$. The number of pores of length $n\Delta x$ is given by $NP_n$, where

$$NP_n = N\phi^n(1 - \phi)^2 \qquad (6.14)$$

Similar reasoning yields $NG_n$, the number of grains of length $n\Delta x$

$$NG_n = N\phi^2(1 - \phi)^n \qquad (6.15)$$

Table 6.1 contains expected and computed numbers of pores and grains of various lengths in a sample of 10,000 points across a hypothetical porous material having 25% porosity. Points were generated with a pseudo-random number generator that produces values of $x_i$ uniformly distributed in the interval $0 \leqslant x_i \leqslant 1$. If $x \leqslant \phi$, the point i is considered to be in a pore; otherwise, in a grain. Figure 6.7 shows typical traces of the pore-grain succession corresponding to such a model. Figure 6.8 is a two-dimensional image of a mathematically synthesized porous material created by this random process. Each point in a 1,024 x 1,024 array is the center of a square area that has been photographically exposed and

TABLE 6.1 Theoretically predicted and actually observed frequencies of pore and grain lengths in sequence of 10,000 points generated at 25% porosity

| Grain-Length Histogram | | | Pore-Length Histogram | | |
|---|---|---|---|---|---|
| Length | Observed | Calculated | Length | Observed | Calculated |
| 1 | 1,482 | 1,468.75 | 1 | 1,437 | 1,406.25 |
| 2 | 361 | 351.56 | 2 | 331 | 351.56 |
| 3 | 221 | 263.67 | 3 | 86 | 87.89 |
| 4 | 216 | 197.75 | 4 | 29 | 21.97 |
| 5 | 149 | 148.32 | 5 | 6 | 5.49 |
| 6 | 136 | 111.24 | 6 | 1 | 1.37 |
| 7 | 108 | 83.43 | 7 | 0 | 0.34 |
| 8 | 55 | 62.57 | 8 | 0 | 0.09 |
| 9 | 38 | 46.93 | 9 | 0 | 0.02 |
| 10 | 35 | 35.20 | 10 | 0 | 0.01 |
| 11 | 18 | 26.40 | 11 | 0 | 0.00 |
| 12 | 13 | 19.80 | 12 | 0 | 0.00 |
| 13 | 8 | 14.85 | 13 | 0 | 0.00 |
| 14 | 8 | 11.14 | 14 | 0 | 0.00 |
| 15 | 17 | 8.35 | 15 | 0 | 0.00 |
| 16 | 4 | 6.26 | 16 | 0 | 0.00 |
| 17 | 4 | 4.70 | 17 | 0 | 0.00 |
| 18 | 2 | 3.52 | 18 | 0 | 0.00 |
| 19 | 5 | 2.64 | 19 | 0 | 0.00 |
| 20 | 2 | 1.98 | 20 | 0 | 0.00 |
| 21 | 3 | 1.49 | 21 | 0 | 0.00 |
| 22 | 2 | 1.11 | 22 | 0 | 0.00 |
| 23 | 1 | 0.84 | 23 | 0 | 0.00 |
| 24 | 0 | 0.63 | 24 | 0 | 0.00 |
| 25 | 0 | 0.47 | 25 | 0 | 0.00 |
| 26 | 2 | 0. | 26 | 0 | 0.00 |
| Totals | 1,890 | 1,873.59 | | 1,890 | 1,875.00 |

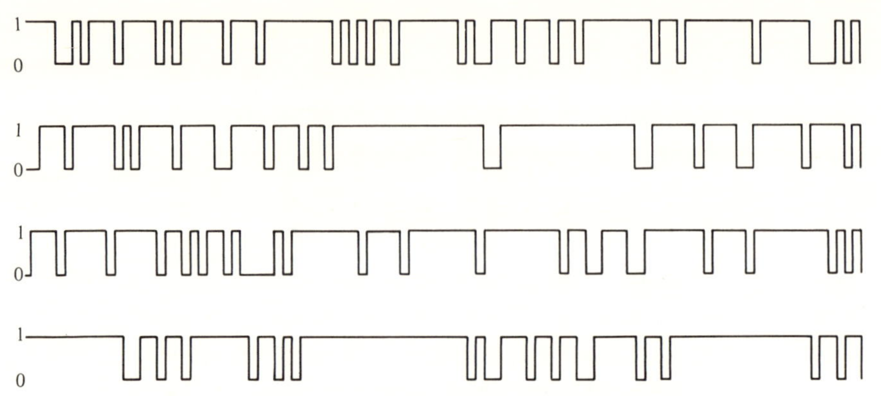


FIGURE 6.7 Square-wave function similar to that shown in Figure 6.1, created by independent random stochastic model. Probability of pore is equal to porosity, 25% in this experiment.

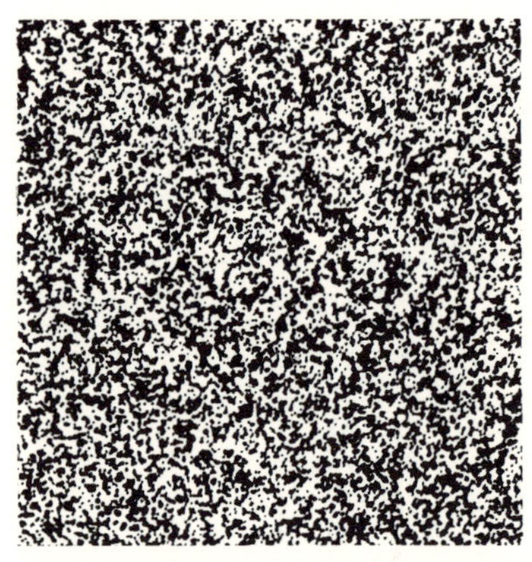

FIGURE 6.8 Two-dimensional image of synthesized porous material created by independent random stochastic model that is direct extension of that used to create Figure 6.7. Compare with actual sandstones in Figure 6.2.

darkened if $x_{i,j} \leqslant \phi$. The image has a striking resemblance to a uniform fine-grain sandstone.

Mathematical representations of porous material such as these also may serve as the basis for models of fluid behavior in porous rocks. For example, M. Y. Joshi (personal communication, 1972) has proposed that tortuosity can be modeled by a process similar to a random walk.

Tortuosity T is defined as $(La/L)^2$ (Carman, 1937; Rose and Bruce, 1949; Wyllie and Rose, 1950) or as (La/L) (Winsauer and others, 1952), where La is the average length of fluid-flow paths through a linear-flow system of length L. The definition used here is $T = (La/L)^2$. Figure 6.9 represents a model in which pores and grains are distinguished only at nodes of a rectangular grid. No specifications of state are made at other locations either perpendicular or parallel to the flow direction.

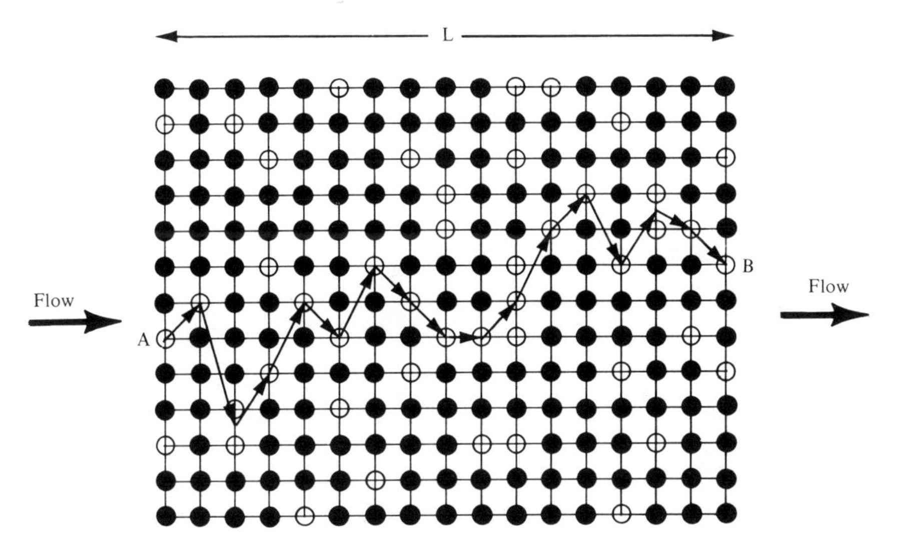


FIGURE 6.9 Hypothetical flow from node to node through a two-dimensional network where pores are represented by open circles and grains by black spots. Model is two-dimensional random walk.

Assume that a particular starting at A travels to the next forward line of nodes by the shortest path; that is, to the nearest pore in the next line. Where two pore nodes in the next line are equidistant from a point in the previous line, and if they are contiguous, a particle goes to a point between them. Where two pore nodes are equidistant but not contiguous, a random choice is made between the two. The total path length is computed by simple geometric relations. This process can be directly extended to three dimensions.

Whether a particular point ($X_{i,j,k}$) represents a pore or grain is determined by a uniform random-number generator. A 500-step random walk was used in the tortuosity model. Walks were repeated five times and mean tortuosity and standard deviation of tortuosity were computed. Porosities were chosen to range from 5 to 80%. Results are listed in Table 6.2 and shown graphically in Figure 6.10. A linear regression of log T versus log $\phi$ for these data gives the following relation between tortuosity and porosity:

$$T = \frac{0.973}{\phi^{0.293}} \tag{6.16}$$

The formation factor F is defined as the electrical resistivity of a porous rock saturated with brine, divided by the resistivity of the brine itself. Archie (1942) formulated the relation between porosity and formation factor as

TABLE 6.2 Tortuosity as function of porosity for simulated three-dimensional porous material

| Porosity (%) | Mean La/L | Std. Dev. of La/L | Tortuosity $T = (La/L)^2$ |
|---|---|---|---|
| 5 | 2.481 | 0.0785 | 6.16 |
| 8 | 2.072 | 0.0616 | 4.29 |
| 10 | 1.923 | 0.0380 | 3.70 |
| 12 | 1.771 | 0.0128 | 3.14 |
| 15 | 1.679 | 0.0569 | 2.82 |
| 20 | 1.516 | 0.0155 | 2.30 |
| 25 | 1.417 | 0.0260 | 2.01 |
| 30 | 1.363 | 0.0197 | 1.86 |
| 40 | 1.285 | 0.0197 | 1.65 |
| 50 | 1.218 | 0.0133 | 1.48 |
| 80 | 1.082 | 0.0072 | 1.17 |

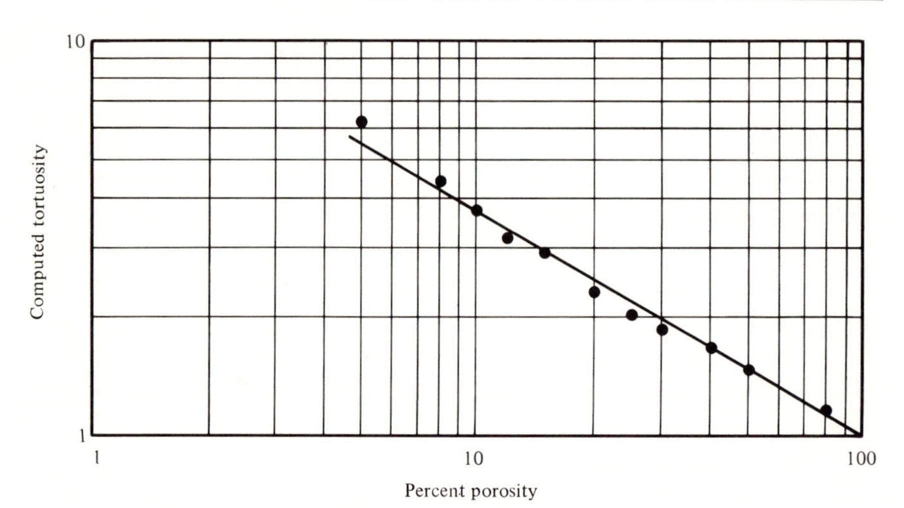


FIGURE 6.10 Linear regression of log tortuosity versus log porosity for random walk models of pore structures having different porosities.

$$F = \frac{1}{\phi^m} \tag{6.17}$$

where F = formation factor

$\phi$ = porosity

The relationship among formation factor, tortuosity, and porosity may be inferred from the analogy between electrical and hydraulic tortuosities (Pirson, 1963):

$$F = T^{1/2}/\phi \tag{6.18}$$

Combination of these two equations yields

$$T = \frac{1}{\phi^{2m-2}} = \frac{1}{\phi^{n}} \tag{6.19}$$

if n is defined as a constant equal to 2m - 2. Results of the least-squares analysis of the model data, given in Equation 6.16, agree closely with the relation $T = 1/\phi^{0.30}$ found experimentally by Archie for unconsolidated sands.

The same random model was used to create both the two-dimensional pore-grain image shown in Figure 6.8 and the measures of three-dimensional tortuosity. The realistic behavior of the flow model and the equally realistic appearance of the image substantiate the idea that simple stochastic processes can serve as models of porous sandstones.

## THE UNIQUENESS PROBLEM

The "true causes" of a real physical system cannot be determined by mathematical modeling because of lack of uniqueness in the algorithms and assumptions that will lead to acceptable simulations. This problem exists with both deterministic and stochastic models. For example, if the joint probability density functions of a stochastic process together form a multivariate normal distribution, then the process is described completely by its multivariate mean, variance, and autocovariance function. However, different nornormal processes can possess the same autocovariance. Thus, the autocovariance function is a unique descriptor only for certain processes. At present, it is not known how significant this problem is in the simulation of pore structure. A similar dilemma is encountered in the interpretation of gravity anomalies, as an observed anomaly may result from various combinations of depth of burial and density contrasts. Peripheral information must be used to choose among alternative interpretations.

Lack of a unique relation between the autocovariance function and a general stochastic model should not preclude the use of random processes in modeling porous material. A realistic simulation is possible using random variables, although a mechanistic interpretation of each variable in the model is not always possible.

## CONCLUSIONS

The need for a new approach to the description of sedimentary rocks is necessary if the nature of the constituent particles are to be related to such properties as porosity, permeability, and formation factor other than empirically. Stochastic processes based on the theory of random variables are suggested for the realistic modeling of the pore-grain structure of clastic sediments. If clastic sediments are realizations of stochastic processes, the autocovariance function of the pore-grain succession, when extended to three dimensions, should contain all information implied in Griffiths' function f(m,s,sh,o,p), except for mineralogy. A stochastic process with a theoretical autocovariance function of the same form as the three-dimensional autocovariance function measured on a natural porous material will permit realistic modeling of all physical properties that depend upon pore-grain geometry.

## ACKNOWLEDGMENT

This research was sponsored in part by the American Petroleum Institute as API Research Project 131.

## REFERENCES

Archie, G. E., 1942, The electrical resistivity log as an aid in determining some reservoir characteristics: Trans. AIME, v. 146, p. 54-62.

Blackman, R. B., and Tukey, J. W., 1958, The measurement of power spectra (from the point of view of communications engineering): Dover, New York, 190 p.

Carman, P. C., 1937, Fluid flow through granular beds: Trans. Inst. Chemical Engineering, London, v. 15, p. 150.

Emery, J. R., and Griffiths, J. C., 1954, Reconnaissance investigation into relationships between behavior and petrographic properties of some Mississippian sediments: Pennsylvania State Univ. Min. Ind. Experiment Stat. Bull., No. 62, p. 67-80.

Fara, H. D., and Scheidegger, A. E., 1961, Statistical geometry of porous media: Jour. Geophysical Res., v. 66, no. 10, p. 3279-3284.

Greenkorn, R. A., and Johnson, C. R., 1964, Directional permeability of heterogeneous anisotropic porous media: Trans. SPE, v. 232, pt. 2, p. 124-132.

Griffiths, J. C., 1952, Measurement of the properties of sediments (abs.): Geol. Soc. America Bull., v. 63, no. 12, pt. 2, p. 1256.

Griffiths, J. C., 1961, Measurement of the properties of sediments: Jour. Geology, v. 69, no. 5, p. 487-498.

Griffiths, J. C., 1967, Scientific method in analysis of sediments: McGraw-Hill Book Co., New York, 508 p.

Jenkins, G. M., and Watts, D. G., 1968, Spectral analysis and its applications: Holden-Day, San Francisco, 525 p.

Kahn, J. S., 1956a, The analysis and distribution of the properties of packing in sand-size sediments. 1. On the measurement of packing in sandstones: Jour. Geology, v. 64, no. 4, p. 385-395.

Kahn, J. S., 1956b, Analysis and distribution of packing properties in sand-sized sediments. 2. The distribution of the packing measurements and an example of packing analysis: Jour. Geology, v. 64, no. 6, p. 578-606.

Kanal, L., and Chandrasekaran, B., 1969, Recognition, machine "recognition," and statistical approaches, in Methodologies of pattern recognition: Academic Press, London, p. 317-332.

Maasland, M., 1957, Soil anisotropy and drainage of agricultural lands, in Drainage of agricultural lands: Am. Soc. Agronomy Monograph, v. 7, p. 216-285.

Maasland, M., and Kirkham, D., 1959, Measurement of permeability of triaxially anisotropic soils: Am. Soc. Chemical Engineers, Proc., v. 85 (SH 3 No. 2063), p. 25-34.

Marks, W., 1969, An engineering guide to spectral analysis: Tech. Rept. No. 69-63, Oceanics Inc., Plainview, New Jersey, ONR Contract NONR-3961(00), NR-083-171/6-1-67, Clearinghouse No. AD 698-837, 72 p.

Matheron, G., 1962, Traite de geostatistique appliquee, tome I, methodes generales d'etude des variables regionalisees: Bureau de Recherches Geologiques et Minieres, Mem. 14, 333 p.

Minoura, N., and Conley, C. D., 1971, Technique for impregnating porous rock samples with low-viscosity epoxy resin: Jour. Sed. Pet., v. 41, no. 3, p. 858-861.

Otto, G. H., 1938, The sedimentation unit and its use in field sampling: Jour. Geology, v. 46, no. 4, p. 569-582.

Pincus, H. J., 1969, Sensitivity of optical data processing to changes in fabric. Part I. geometric patterns; Part II. standardized grain size patterns; Part III. rock fabrics: Int. Jour. Rock Mech. Min. Sci., v. 6, p. 259-276.

Pirson, S. J., 1963, Handbook of well log analysis: Prentice Hall, Inc., Englewood Cliffs, New Jersey, 326 p.

Preston, F. W., and Davis, J. C., in press; Application of optical processors to geological images, in Machine perception of patterns and pictures: Nat. Physical Lab., London.

Roach, S. A., 1968, The theory of random clumping: Methuen and Co., Ltd., London, 94 p.

Rose, W., and Bruce, W. A., 1949, Evaluation of capillary character in petroleum reservoir rock: Trans. AIME, v. 186, p. 127-142.

Shulman, A. R., 1970, Principles of optical data processing for engineers: Goddard Space Flight Center, Greenbelt, Maryland, NASA Tech. Rept. R-327, 122 p.

Tukey, J. W., 1966, An introduction to the calculations of numerical spectrum analysis, in Spectral analysis of time series: John Wiley & Sons, Inc., New York, p. 25-43.

Tukey, J. W., 1970, Some further inputs, in Geostatistics: Plenum Press, New York, p. 163-174.

Winsauer, W. O., Shearin, H. M., Jr., Masson, P. H., and Williams, M., 1952, Resistivity of brine saturated sands in relation to pore geometry: Am. Assoc. Petroleum Geologists Bull., v. 36, no. 2, p. 253-277.

Wyllie, M. R. J., and Rose, W. D., 1950, Some theoretical considerations related to the quantitative evaluation of the physical characteristics of reservoir rock from electrical log data: Trans. AIME, v. 189, p. 105-118.

# Analysis of Volcanic Earthquakes of Asamayama (Japan)

# 7

R. A. Reyment

ABSTRACT

The statistical theory of point processes has been used to analyze a series of observations on volcanic earthquakes derived from the Japanese volcano Asamayama. Preliminary studies of the observations disclose a seeming cyclicity. The earthquakes do not occur in accordance with a Poisson type of process, nor are the intervals between events independently distributed, but there is a tendency for the events to cluster (high coefficient of variation). An analysis of the serial correlation coefficients leads to the rejection of a renewal hypothesis, and specific tests for renewal hypotheses support this conclusion. Analysis of the periodogram shows the existence of significant trend in the data, and specific tests lead to the rejection of the hypothesis of independence between intervals. The analysis of the graph of the logarithmic empirical survivor function shows it to differ from the form expected for exponentially distributed times between events.

## INTRODUCTION

Recently, I published statistical analyses of the eruptive patterns of some active volcanoes (Reyment, 1969), the data on which were extracted from the "Catalogue of the Active Volcanoes of the World." The object of that exercise was to see how well the newly developed theory of stochastic point processes (Cox and Lewis, 1966) could be applied to the treatment of such material. As a result of this study, three Japanese volcanoes - Asamayama (Aramaki, 1963), Asosan, and Kirisimayama were determined to possess a similar pattern of eruption, dominated by trend in the rate of occurrence of events (= eruptions), with an overriding tendency for an increase in the relative frequency of the events.

The observations published in the "Catalogue" are derived from various sources, many of them made by untrained observers. Since 1966, the Japanese Meteorological Agency has been issuing data on the eruptive activity of several volcanoes in the Japanese islands, including the three previously mentioned. The observations on volcanic earthquakes provide a good source of material for testing ideas on natural stochastic processes. The figures on volcanic smoke are far less useful because of the obvious difficulty and arbitrariness involved in their collection.

## ORDER OF ANALYSIS

Analysis of the sequence of observations on volcanic earthquakes of Asamayama was based on the theory of point processes. The approach underlying the treatment of the material was to test by different methods whether the earthquakes could be regarded as approximating to some type of stochastic point process. The theory of point processes was developed originally as a response to the need of being able to examine series of events occurring in continuous unidimensional time. It is well known that many natural processes occur randomly in time. Among those of geologic interest is the behavior of volcanoes and earthquakes; a vast literature is available on the subject.

Various aspects of the activity of volcanoes and the occurrence of earthquakes may be amenable to study by the methods of point process analysis. In my original analysis of volcanologic observations (Reyment, 1969, p. 59), I pointed out that the result of any attempt at using published data on dates of eruptions must be approximate, owing to the variable and unreliable nature of the input material. The information analyzed in this paper has a satisfactory degree of accuracy. It reflects the activity (= liveliness) of a volcano, although it cannot be equated directly with the concept of an actual volcanic eruption. Inasmuch as volcanic earthquakes are of short duration, almost instantaneous, in relation to the times between events, they are ideally suited to the point process model.

A suitable starting point for the analysis of a series of events is, after the preliminary appraisal of the graph of the cumulative number of events, the Poisson distribution. It is a reasonable assumption, as a first approximation, that a natural process would occur randomly in time. As I demonstrated (Reyment, 1969), however, this is not a well-based assumption for volcanoes, and only a few of the examples given in that paper agree with a Poisson model with respect to eruptive histories. The reason for this is that a volcanic outbreak is almost always the result of a complex of components, the optimum behavior of all or most of which may be necessary in order that an eruption be realized. Thus, there is the absolute need to be certain as to what constitutes an eruption and what constitutes persistent activity. This distinction is seldom used by untrained observers. The volcanic earthquakes of the volcanoes covered by the Japanese Volcanological Bulletin are registered automatically, which reduces the human element as a source of observational error (1966, p. 5).

With the Poisson model as a cornerstone, one may branch into the wider field of renewal processes. The intervals between the successive events of a renewal process are independent of each other and identically distributed. Where these intervals are distributed exponentially, one has the special situation of the Poisson process.

For the purposes of the analysis presented in this paper, I employ a convention that adheres to the assumption that an event may have occurred immediately before the starting point of the analyzed sequence, i.e., prior to time $T = 0$. It is assumed also that there is no chance that two or more events can occur simultaneously. This latter assumption is necessary for the statistical theory to apply and it is also a logical rule to have in almost all situations that may be encountered by statisticians in their everyday work. The geologist is less likely to support this point of view, at least with respect to subsidiary volcanologic phenomena, such as volcanic earthquakes.

The order of analysis requires that the data first be tested for trend. If significant trend can be proved to exist in the material, the suite of observations cannot be approximated by a point process. If significant trend cannot be demonstrated to occur, usually one may be able to postulate that the series is stationary. Subsequent analytical interest will center about attempts at establishing whether the successive events are dependent or independent. The serial correlation coefficient plays an important part in this part of the analysis.

Details of the statistical methods used in the study of point processes in geology are given in Reyment (1969).

## ORIGIN OF DATA

The observations treated here were taken from the Volcanological Bulletin of the Japanese Meteorological Agency (1966-1967). The data consist of the exact time of volcanic earthquakes recorded by the seismic station on Asamayama. From these it is possible to construct the required sequence of intervals.

## EARTHQUAKES OF ASAMAYAMA

Owing to the great length of the observational series on the volcanic earthquakes of the volcano, a reflection of its considerable seismic activity, it was necessary to partition the data in order to get the material onto a CDC 3600 computer, a machine with only moderate storage capacity. The irksome necessity of having to partition the data was a blessing in disguise, as it disclosed the unsuspected structural complexities in the sequence of observations.

The results of my original analysis of Asamayama, Asosan, and Kirisimayama (Reyment, 1969) showed that all of the eruptive sequences of these volcanoes display significant trend in their respective rates of occurrences of events. The present analysis shows that the same general behavioral pattern typifies the earthquakes of Asamayama; most of the registered sequence of events shows significant trend, whereas a part of the sequence lacks significant trend. The two patterns are analyzed separately.

### Volcanic Earthquakes Without Significant Trend

Approximately the first 500 shocks of the observational sequence lack significant trend in their rate of occurrence ($U = -0.08$). $U$ is the statistic calculated in the test procedure given in Cox and Lewis (1966, p. 47). Significant trend in the data means that, instead of the rate parameter for a Poisson process model for the sequence being constant over time, the following relationship applies:

$$\lambda(T) = e^{\alpha+\beta T} \quad (7.1)$$

where $T$ is the fixed length of time during which the events have taken place and $\beta$ describes the nature and direction of the trend.

If the methods used to test the Poisson model lead to a rejection of this relationship, some other type of renewal process should be considered. A useful first step in this direction is to study the serial correlation coefficients $\rho_i$ of the series (see Cox and Lewis, 1966, p. 91). For a renewal process, $\mathrm{var}(\tilde{\rho}_i) \sim 1/(n - i)$, where $\tilde{\rho}_i$ is the estimate of the serial correlation coefficient. The test of the hypothesis, $\rho = 0$, is made by means of $\tilde{\rho}_i\sqrt{(n - i)}$, which has approximately a unit normal distribution for large $n$ (say, $n > 100$). If the serial correlation coefficients are not significant, i.e., beyond what is to be expected from the vagaries of chance (Reyment, 1969, p. 64), one may be reasonably certain that the series agrees with a renewal model as far as the criterion of independence of the intervals between successive events is concerned.

The first 20 serial correlation coefficients and the corresponding normal approximations of the first set of 499 observations are given in Table 7.1. Six of these coefficients are significant at the 5% level; moreover, 18 of 20 coefficients are positive. Both these facts lessen the likelihood of the intervals between events being independently distributed. For the total section analyzed, 13 of 100 correlation coefficients are significant, which is far more than is allowed by the 95% level of risk (this would be 5 out of 100).

The graph of the logarithmic empirical survivor function is shown in Figure 7.1. For exponentially distributed intervals between successive events, the line should be linear over almost its entire course. The line in Figure 7.1 has a fairly well-developed regular convex-downward shape, indicating departure from the exponential distribution.

Despite the negative evidence yielded by the analysis of the serial correlation coefficients, it is prudent to make specific tests for Poisson processes as part of the complete analysis. The first matter of interest concerns the size of the coefficient of variation for the intervals between events, which is 1.48 and in excess of the theoretical value of unity for a Poisson process.

The second step is formed by the calculation of some distribution-free tests for Poisson processes, namely, the one- and two-sided Kolmogorov statistics ($D_n^+$, $D_n^-$, and $D_n$) and the Anderson-Darling statistic $W_n^2$, using transformed data to increase the power of the tests (Lewis, 1967, p. 210). The results of these tests are given in Table 7.2. The asterisks indicate rejection of the Poisson hypothesis at the 5% level at least. Inspection of these values shows that the Poisson hypothesis is strongly rejected.

The analysis of the periodogram estimate of the power spectrum also is instructive (Cox and Lewis, 1966, p. 98). Deviating estimates of the power spectrum show trends in their values that may be determined by means of the statistic U, as used for identifying trend in the original observational sequences. The test for independence is to find out if the periodogram for successive values agrees with a Poisson process. The results are definite. Not only is there a significant trend in the sequence, but the values of the Kolmogorov-Smirnov and Anderson-Darling tests are highly significant (e.g., $D_n = 1.94$ and $W_n^2 = 5.56$) as well.

A preliminary graphic analysis of the series of events shows them to have a tendency to form clusters, which is substantiated by the high coefficient of variation. A renewal model is rejected by an analysis of the serial correlation coefficients and by specific tests for a Poisson process, as well as the analysis of the periodogram.

TABLE 7.1 Serial correlation coefficients for Asamayama

| $i$ | 1 | 2 | 3 | 4 | 5 | 6 | 7 | 8 | 9 | 10 |
|---|---|---|---|---|---|---|---|---|---|---|
| $\tilde{\rho}_i$ | 0.101[a] | 0.159[a] | 0.119[a] | 0.068 | 0.015 | 0.014 | 0.063 | 0.089[a] | 0.018 | 0.014 |
| $\tilde{\rho}_i\sqrt{(n-i)}$ | 2.230 | 3.514 | 2.626 | 1.496 | 0.332 | 0.319 | 1.379 | 1.960 | 0.398 | 0.307 |
| $i$ | 11 | 12 | 13 | 14 | 15 | 16 | 17 | 18 | 19 | 20 |
| $\tilde{\rho}_i$ | 0.024 | -0.004 | 0.036 | 0.118[a] | 0.142[a] | 0.007 | 0.036 | 0.010 | 0.022 | -0.041 |
| $\tilde{\rho}_i\sqrt{(n-i)}$ | 0.521 | -0.093 | 0.791 | 2.570 | 3.103 | 0.152 | 0.785 | 0.210 | 0.468 | -0.892 |

[a]Denotes significance at 5% level.

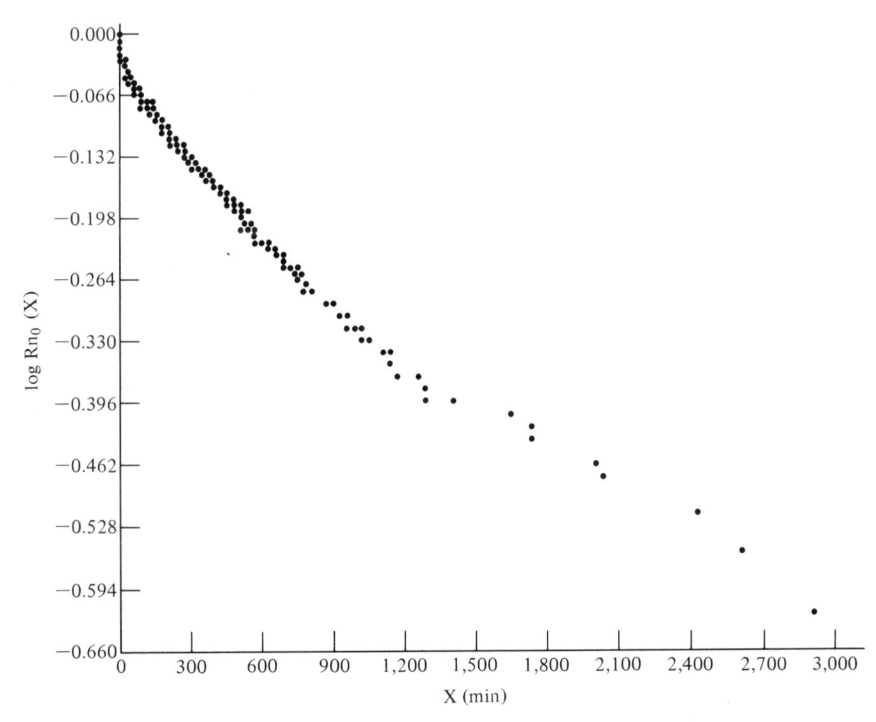


FIGURE 6.1 Graph of logarithmic empirical survivor function for intervals between shocks for trend-free phase in earthquake history of Asamayama.

TABLE 7.2 Results of tests for Poisson model

| $D_n^+$ | $D_n^-$ | $D_n$ | $W_n^2$ |
|---|---|---|---|
| 264.4[a] | -0.01 | 264.4[a] | 69.3[a] |

[a]Denotes significance at 5% level.

Volcanic Earthquakes with Significant Trend

The analysis of the main part of the sequence of observations following directly on the set just described, without trend, showed an important deviation from the latter in that the entire run displays trend, albeit with significant directional differences. The series has an oscillatory character, as it were.

The values for the statistic U for the arbitrary sets of sequentially ordered observations are presented in Table 7.3. These results indicate clearly that the direction of trend fluctuates strongly. Thus, the intervals between earthquake shocks in the sequence of observations available (15 months) are trend-free to begin with, although they do not conform with any type of renewal process (Reyment, 1969, p. 75). They change

TABLE 7.3 Trend results for arbitrary sets of consecutive observations

| | n | U |
|---|---|---|
| 1 | 492 | -6.13 |
| 2 | 499 | 3.14 |
| 3 | 500 | 2.53 |
| 4 | 495 | -3.65 |
| 5 | 496 | -2.94 |

character by developing a significant trend with $\beta > 0$ in Equation 7.1, which indicates a decrease in the rate of occurrence of events; thereafter, there is a long period (about 6 months) during which an opposite trend occurs, with $\beta > 0$, followed by a new phase with $\beta < 0$. Although these swings in direction can be recognized by inspection of the graph of the cumulative number of events against time, they are not immediately obvious, owing to the great number of observations, without some assistance from hindsight.

The total analysis was based on a record of almost 3,000 shocks over a 15-month period. During the same period, the volcano Asosan experienced only 144 shocks, showing trend with $\beta > 0$ ($U = 3.07$).

Obviously, the zones between switchovers from negative to positive trending must yield values of U of around 0 or, at least, below the level of significance. Inspection of the data discloses, however, that these switchover zones are narrow, far narrower than the stretch of observations represented by the first set of trend-free data.

## CONCLUSIONS

The volcano Asamayama displays trend in the rate of occurrence of events, not only in the pattern of its major eruptions but also in a subsidiary phenomenon of its activity, notably, the volcanic earthquakes. The earthquake activity during the period analyzed here passed from a phase of trendless behavior, characterized by statistical dependence between successive events, to a phase of successively diminishing activity, followed by a phase in which the rate of occurrence of shocks increased. The final part of the sequence studied is marked by a return to a phase of lessening in the frequency of shocks.

## ACKNOWLEDGMENTS

The cost of the computer calculations was defrayed by Grant 104104 of the University of Uppsala. The computing was done on the CDC 3600 of the

Computer Centre of the University (Datacentralen) using a slightly modified version of the original program (Reyment, 1969).

## REFERENCES*

Aramaki, S., 1963, Geology of Asama volcano: Jour. Fac. Sci. Univ. Tokyo, Sec. II, v. 14, p. 229-443.

Cox, D. R., and Lewis, P. A. W., 1966, The statistical analysis of series of events: Methuens Monographs on Applied Probability and Statistics, London, 285 p.

Lewis, P. A. W., 1967, A computer program for the statistical analysis of series of events: IBM Systems Jour., v. 5, no. 4, p. 202-225.

Reyment, R. A., 1969, Statistical analysis of some volcanological data regarded as series of point events: Pure and Applied Geophys., v. 74, p. 57-77.

Volcanological Bull., 1966 to 1967: Japanese Meteorological Agency, v. 5, nos. 1-4; v. 6, nos. 1-4.

*A more comprehensive bibliography is given in Reyment (1969).

# 8

# Stratigraphic Implications of Random Sedimentation

W. Schwarzacher

ABSTRACT

Stratigraphic time can be measured by a variety of methods. In its original meaning, stratigraphy takes its time scale from the rates of sedimentation. However, sedimentation rates are not constant, and some simple random processes are used to describe this situation. Having decided on such a sedimentation model, one can examine its stratigraphic behavior.

## INTRODUCTION

A number of stochastic sedimentation models have been introduced in recent years, but their impact on stratigraphy, in general, has been limited. Although many reasons can be given, there are perhaps two main causes. The new concepts have been incompletely understood and uncritically applied, usually under the illusion that they are a ready-made tool for the analysis of sections. More important, however, is the fact that most sedimentation models are simple and primitive from a geologist's point of view.

The simplicity of models is unavoidable to a certain extent, because our knowledge of sedimentation processes is incomplete, and, without basic facts, it is impossible to develop more complicated models. In addition, the models must be reasonably simple in order that they be understood.

The problems and examples discussed in this paper arose directly from a study of a carbonate basin in the northwest of Ireland. However, the models are kept simple at this stage of development and no particular importance is attached to the regional setting of the basin. Thus, they should be of general interest to the stratigrapher.

## SOME SIMPLE MODELS OF SEDIMENTATION

Consider a basin in which some sediment is deposited. It is assumed that each locality in this basin produces its own sediment and that this sedimentation is constant throughout the basin's history. Certain events occur, recorded by marker horizons in the sediment, and it is assumed that these events do not affect the sedimentation process (Figure 8.1). It is assumed also that the time events (T) are recorded over the entire basin. The thickness of sediment between two markers will be called a bed for the sake of simplicity. The thickness of the beds is clearly proportional to the time they record; furthermore, the thickness of individual beds is constant over the entire basin.

In order to develop this idea further, it may be assumed that sedimentation is disturbed by random interruptions. As an example, one might think of storms that periodically destroy the carbonate-producing organisms. It is assumed also that each interruption consists of a constant length of time during which no sedimentation occurs. The sequence of such "storms" is unpredictable and is represented by a Poisson process.

This model is called the "interrupted sedimentation model," shown in Figure 8.2A. The sedimentation is retarded by N(t) times the loss caused by periods of nondeposition. N(t) is the number of interruptions and is a random variable. The interrupted sedimentation model in this simple form is exactly equivalent to a second model that assumes discontinuous sedimentation. Sedimentation moves a step at each interruption; between

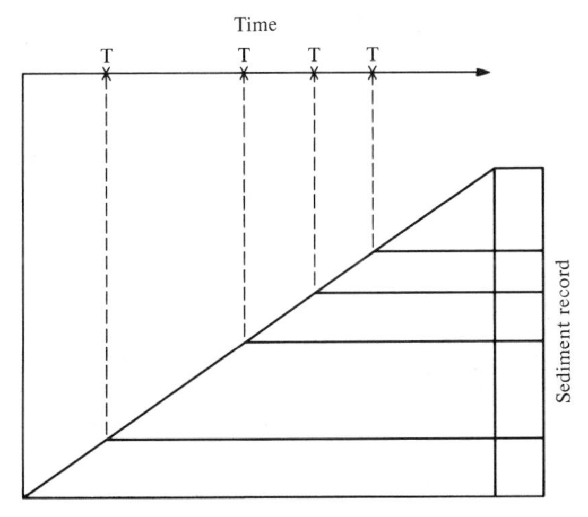


FIGURE 8.1 Simple deterministic sedimentation model.

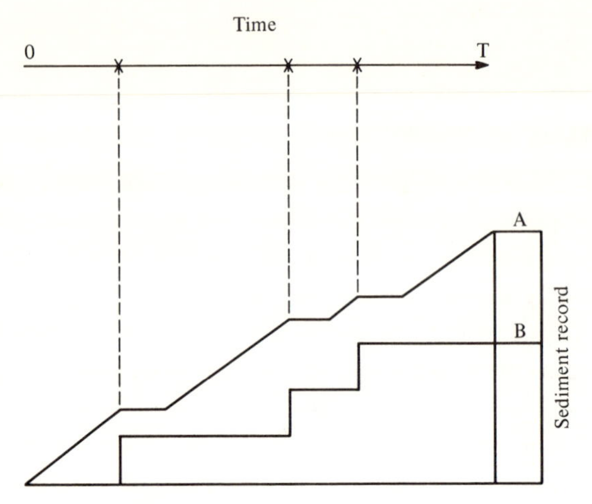


FIGURE 8.2 Simple stochastic sedimentation models. (A) Interrupted sedimentation; (B) Step sedimentation.

storms no sedimentation takes place. The second model is shown in Figure 8.2B. The second model seems to be more realistic for the particular limestones examined in the northwest of Ireland. Almost steplike sedimentation is indicated by the in situ preservation of the benthonic fauna as well as other ecological and petrographic evidence. The thickness of the step is controlled mainly by the biological relief present at the time on the seafloor. Whether one is actually dealing with storms or other climatic or biologic incidents is not relevant, and the term "storm" is retained only for simplicity of language.

The step model could be refined further by postulating that each storm is followed by a recovery time during which deposition is impossible. No deposition occurs if a storm happens within this recovery time. It can be assumed either that such a storm is completely inactive (type 1) or, alternatively, that the storm may extend the dead time (type 2). In this model one recognizes the electronic counter that has been studied extensively for type 1 and 2 problems (Feller, 1957; Bharucha-Reid, 1960). Similar assumptions can be made for the interrupted sedimentation model, and with the use of counter terminology the main interest in this situation would be the length of the blocked period. Of course, the step model and the interrupted sedimentation model can be combined to treat more complex processes.

It is evident that the lateral variation of bed thickness in these models will depend on the area over which the storms operate. Consider the extreme situation in which each locality is independent and has its own sequence of storms in contrast to the situation in which storms occur simultaneously over the entire basin. In the first situation, lateral variation becomes identical to vertical bed-thickness variation, but such a model is not realistic. The independence would lead to sections in which, for example, the average thickness of each bed is the same, and this is contrary to observations. It is more likely that the interruptions are active over large areas or even the entire basin, particularly if not too large. It is possible, however, that each locality reacts differently and the amount of sediment laid down in each storm is, in itself, a random variable; the recovery time following each storm also may differ according to some probability law.

The difference between independent and partly dependent lateral variation can be seen in the two simulated sections of Figure 8.3. In Figure 8.3A the vertical as well as the lateral thicknesses are Poisson, but each individual bed follows a special Erlang distribution. It is interesting to note that in the second situation, sections could be correlated by matching their thicknesses, which is not possible in the independent example.

## THE SEDIMENTATION MODEL AS A RENEWAL PROCESS

For the mathematical analysis, it is convenient to consider the outlined models as renewal processes. The history of sedimentation is regarded as a sequence of periods in which deposition can take place (counter open) followed by intervals in which no deposition may occur (counter blocked). A sedimentation step, or deposition gap, occurs at the beginning of each blocked period, which, in counter terms, would be called a registration. Each registration is a renewal point in the sequence and the first important question concerns the number of renewals in a given time

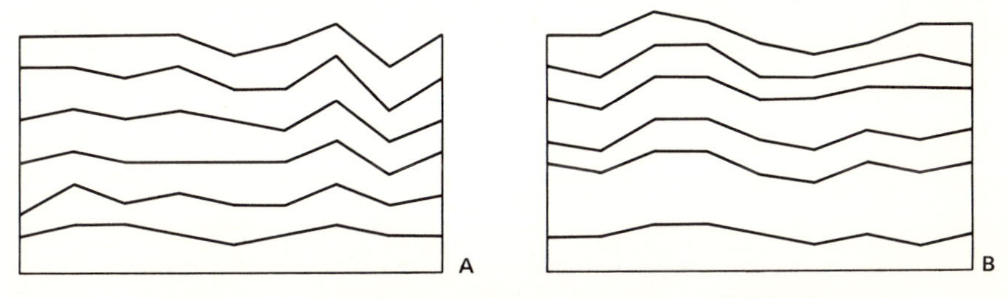


FIGURE 8.3 Simulated geologic sections. (A) Both vertical and lateral thicknesses are Poisson distributed. (B) Vertical sequence is Poisson distributed and has thicknesses in different localities following special Erlang distribution.

interval. The theory of renewals has been fully treated in a monograph by Cox (1962).

Using the notation of Cox (1962, p. 33) the time to the rth renewal $S_r$ has the cumulative distribution $K_r(t)$. The probability of obtaining exactly N renewals in the time interval is given by

$$P(N(t) = r) = K_r(t) - K_{r+1}(t) \tag{8.1}$$

The step model of sedimentation may serve as an example. Let it be assumed that the storms occur at random and follow a Poisson process of rate $\rho$. Each storm is followed by a recovery time that is constant. If the process starts during a period of sedimentation, all open intervals $t_1''$, $t_2''$, . . . (see Figure 8.4) are forward renewal times and therefore are identically exponentially distributed. The sum $S_r$ in time t - r is given by the r-fold convolution of the exponential intervals, which is the special Erlang distribution of index r.

Using the cumulative form of this distribution (Cox, 1962, p. 20), we can enter into Equation 8.1 and see that

$$P(N(t) = r) = \frac{\rho^r (t - rT)^r}{r!} e^{-\rho(t-rT)}$$

Again this is a Poisson distribution and represents a well-known result of the theory of type 1 counters (cf. Bharucha-Reid, 1960). A variable step might be introduced next, and if the recovery time is assumed to be 0 for simplicity, then the probability of obtaining N steps in the time T is given by the compound Poisson distribution

$$P(\theta=x, T=t) = e^{-\rho t} \sum_{0}^{\infty} \frac{(\rho t)}{n!} f^{*n}_{(x)}$$

in which f(x) is the distribution of steps. All models discussed in this paper lead to distributions of this type.

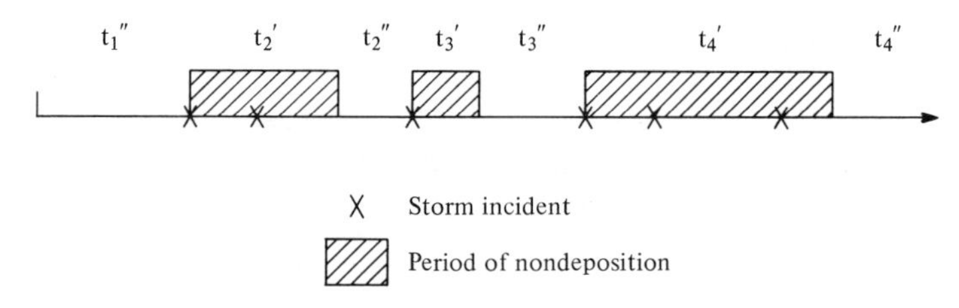


FIGURE 8.4 Sedimentation as renewal process.

One may take the special situation in which f(x) is exponential with density $\lambda$, and if it is assumed that each time interval begins with a depositional step, then Equation 8.2 can be written as

$$P(x,t) = \lambda e^{-(\rho t+\lambda x)} \sum_{0}^{\infty} \frac{(\lambda \rho x t)^n}{(n!)^2}$$

Alternatively, the distribution may be expressed by a Bessel function:

$$P(x,t) = \lambda e^{-(\rho t+\lambda x)} J_0(2i\sqrt{\lambda \rho x t})$$

This model also gives a clear answer to the problem of lateral variation. The thickness distribution, which is measured on a single bed in different localities, will be the special Erlang distribution, with the number of stages corresponding to the number of storms recorded by the bed. From such measurements, $\lambda$ could be estimated and, at least in theory, the model could be checked for consistency.

The interrupted sedimentation model leads to similar results. Referring again to Figure 8.4, and using counter terminology, one can see that the interval between two registrations is given by the convolution of the blocked and open time. The open time is the forward renewal time and therefore is exponentially distributed with the storm density $\rho$. It is assumed that the blocked time also is exponentially distributed with density $\rho$. Entering Equation 8.1, we can write the Laplace transform

$$P^*(n,s) = \frac{1}{s}\left[\left(\frac{\rho\lambda}{(\rho+s)(\lambda+s)}\right)^n - \left(\frac{\rho\lambda}{(\rho+s)(\lambda+s)}\right)^{n+1}\right]$$

The inversion of this transform is difficult but the relation becomes simple in the special situation when $\rho = \lambda$, and the probability for N registrations can be written as

$$P(n,t) = \frac{(\rho t)^{2n}}{(2n)!} e^{-\rho t}$$

Again, this is the Poisson distribution with the dummy variable replaced by 2n. Remembering that the amount of sediment deposited after each recorded storm is proportional to the open time that is exponential, one can see that the bed-thickness distribution again must be a mixed Poisson special Erlang distribution. This model is not capable of predicting the lateral thickness variation of a single bed because, although the number of storms in each bed is constant, the number of effective storms is variable.

The variable step model can be made more realistic by a simple modification. Both observation and theoretical consideration suggest that the steps are probably Gamma distributed (Schwarzacher, 1972). The sedimentation process therefore may be written as

$$P(x,t) = \lambda e^{-(\rho t+\lambda x)} \Sigma \frac{(\lambda^a x^a \rho t)^n}{n!(an)!}$$

in which a and $\lambda$ are the parameters of the step distribution. The step distribution becomes asymptotically normal when a is large. If both a and $\lambda$ increase, the variance decreases until all the probability is concentrated around the mean value of the step distribution. In this limiting situation, the thickness distribution once more becomes discrete (Poisson); transitional to this, however, are polymodal distributions with peaks corresponding to each storm that has been registered (see later examples).

A similar model was developed previously (Schwarzacher, 1972) by using a semi-Markov chain. A two-state Markov matrix (representing deposition and nondeposition and operated at equal time intervals) is used to decide whether a storm occurs. Deposition takes place whenever a storm occurs, and the amount of deposition is determined by the waiting time in the semi-Markov process. If this waiting time has the special Erlang distribution, then the model describes the variable-step sedimentation as before.

## THE PROBLEM OF TIME STRATIGRAPHY

The basic problem of stratigraphy is determining time values that correspond to the measured thicknesses. The stochastic sedimentation model, as discussed, provides a conditional probability, which may be written as

$$P(\theta=x|T = t) \tag{8.2}$$

This probability links thickness $\theta$ and time T and, in a manner of speaking, provides the rules by which sedimentation operates. It is assumed that a model has been selected that gives a satisfactory description of the physical system. The remaining problem can be seen best by using a simple example.

Let the sedimentation be represented by a fixed-step model that operates at intervals given by the Poisson process of density $\rho$. The conditional probability is, therefore, the Poisson distribution

$$P(\theta=x|T=t) = e^{-\rho t} \frac{(\rho t)^x}{x!} \qquad x = 1, 2, \ldots \tag{8.3}$$

Let the time interval T that marks the individual beds in the sequence be distributed as

$$P(T=t) = \lambda e^{-\lambda t} = q(t) \tag{8.4}$$

Equations 8.3 and 8.4 permit the complete joint probability $P(\theta,t) = P(\theta|t) \cdot q(t)$ to be written in the form

| x | t = 1 | t = 2 | t = 3 |
|---|---|---|---|
| 0 | $\lambda e^{-(\lambda+\rho)}$ | $\lambda e^{-2(\lambda+\rho)}$ | $\lambda e^{-3(\lambda+\rho)}$ |
| 1 | $\lambda\rho e^{-(\lambda+\rho)}$ | | |
| 2 | $\frac{(\lambda\rho)^2}{2!} e^{-(\lambda+\rho)}$ | | |
| 3 | | | |

Integration of each row gives the marginal distribution

$$P(\theta=x) = \frac{\lambda}{\lambda+\rho} \left(\frac{\rho}{\lambda+\rho}\right)^x = f(\theta) \tag{8.5}$$

and summing the columns leads to the second marginal distribution q(t).

If this is a real geologic example, only two types of measurement can be made. A single stratigraphic section in one locality will provide the marginal distribution $f(\theta)$. Alternatively, a single bed can be chosen and its thickness variation measured over a lateral range, in which instance the time interval is constant by definition. An estimate of q(t) can be obtained, provided that the thickness variation from locality to locality is stochastically independent.

As has been mentioned, the assumption of independent variation is unrealistic and is excluded from our model by the condition that storms are active simultaneously over the entire basin. If the variation from locality to locality becomes dependent, it can no longer provide uncontaminated information about q(t). Measurements only provide $f(\theta)$ and the sedimentation model provides $p(x|t)$ but the joint probability $p(x,t)$ is needed to obtain q(t).

One might proceed in two directions under these circumstances. Additional models could be used to provide a theory of lateral variation, an approach taken by Jacod and Joathon (1972). The main difficulty is that little is known as yet about this variation, and Jacod and Joathon's assumptions of simple geometric shapes (such as lenses for sedimentary beds) provide a purely deterministic model. When more is known about the processes involved, this will be an interesting field of theoretical stratigraphy.

An alternative approach is to make assumptions about $q(t)$, that is, the distribution of events in time. Two limiting situations are of particular geologic interest. The time intervals leading to bed formation may be distributed completely at random, as indicated by Equation 8.4, or, in complete contrast, the time intervals may be constant. The latter is realized in varved sediments in which one bed represents the time interval of precisely one year, and this hypothesis also can be made in investigations of cyclic sedimentation. The stratigraphic problem causes no difficulty if T is constant because the thickness distribution $f(\theta)$ is obtained directly from the sedimentation model by setting t constant.

In order to examine the question of randomly distributed time intervals, the previous example again can be considered as a renewal process. The thickness of beds in the constant-step model is proportional to the number of steps or renewals $N(t)$. The time interval $(0,T)$, through which renewals are counted, is a random variable with the distribution $q(t)$ (Equation 8.4). Following Cox (1962, p. 37), one can write the generating function of $N(t)$ as

$$G(t,\xi) = \sum_{T=1}^{\infty} \xi^{r} P(N(t) = r)$$

which has the Laplace transform

$$G^*(s,\xi) = \int_0^{\infty} G(t,\xi)e^{-st}dt \tag{8.6}$$

The generating function $G(\xi)$ for renewals $N = 1, 2, \ldots,$ within the random time T is

$$G(\xi) = \int_0^{\infty} G(t,\xi)q(t)dt = \lambda \int_0^{\infty} G(t,\xi)e^{-\lambda t}dt$$

Comparison of this with Equation 8.6 shows that

$$G(\xi) = \lambda G^*(s,\xi)_{s=\lambda}$$

Using Cox's Equation (3.2,4) (1962, p. 37) for the ordinary renewal process, one can write the generating function:

$$G(\xi) = \frac{1 - f^*(\lambda)}{1 - \xi f^*(\lambda)} \tag{8.7}$$

in which $f^*(s)$ is the transform of the time between renewals. If $f^*(s) = \rho/(s + \rho)$, then Equation 8.7 can be written as

$$G(\xi) = \frac{\lambda}{\lambda - \rho(1 - \xi)}$$

which can be expanded to

$$\frac{\lambda}{\lambda + \rho} \sum_{r=0}^{\infty} \left(\frac{\rho s}{\lambda + \rho}\right)^r$$

and therefore

$$P(N = r) = \frac{\lambda}{\lambda + \rho} \left(\frac{\rho}{\lambda + \rho}\right)^r$$

Of course, the result is identical to the one obtained before by direct methods (Equation 8.6). Equation 8.7 is more general, however, and it may be seen that whichever renewal process is substituted, the expression always represents the sum of a geometric expansion. The important geologic result that follows is that if the time input is independently random distributed, the output (in terms of number of steps in the stratigraphic section) is an independent random sequence. Furthermore, it is easily shown that whichever distribution of step thickness is adopted, the thickness distribution $f(\theta)$ will be exponential. This is always true as long as there is a time-thickness relationship in existence.

Any other distribution may exist between the two extreme examples of T being either constant or exponentially distributed. It should be possible to approximate most of these unknown distributions with some special Erlang distribution with the parameters a and $\lambda$. Mixing of the Poisson distribution with the special Erlang distribution and the use of the same methods as in the previous example lead to the negative binomial distribution

$$P(N = r) = \binom{-a}{r} \left(\frac{\rho}{\rho + \lambda}\right)^a \left(- \frac{\lambda}{\rho + \lambda}\right)^r$$

Writing for $\rho/(\rho + \lambda) = K$ and $\lambda/(\rho + \lambda) = q$, one may summarize the stratigraphic problem as follows:

| Distribution of time events $q(t)$ | Distribution of number of renewals $P(N)$ | Distribution of thickness $f(\theta)$ |
|---|---|---|
| T = Constant | $\frac{(c\rho)^r}{r!} e^{-\rho c}$ | Compound Poisson model |
| $q(t) = \lambda \frac{(\lambda t)^{a-1}}{(a-1)!} e^{-\lambda t}$ | $\binom{-a}{r} K^q (-q)^r$ | Compound Poisson |
| $q(t) = \lambda e^{-\lambda t}$ | $qK^r$ | Exponential |

## APPLICATION OF THE THEORY TO STRATIGRAPHIC PROBLEMS

Whether the theory developed has any application to real stratigraphic problems depends on how much is expected from this simplified model. It has been shown that the relation between time and sedimentation can be examined only by making assumptions about the sedimentation process, because the joint probability $P(\theta,t)$ cannot be calculated. This is true under the most favorable conditions, even if both of the marginal conditions $f(\theta)$ and $q(t)$ are known. Therefore, the sedimentation process must be introduced as a hypothesis.

It is reasonable to ask whether such a hypothesis can be tested and, indeed, the sedimentologist may expect that by trying different sedimentation models he may learn something about the sedimentation process itself.

No statistics have been attempted in this paper, but it is obvious that the prospect of testing various models is not promising. All the sedimentation processes that have been considered lead to compound Poisson distributions with at least two unknown parameters; further uncertainties are added if nothing is known about the distribution of time events. Furthermore, if the poor quality of geologic data in general is considered, it becomes unlikely that one can discriminate between a series of related models. At best, one may be able to establish that a chosen model does not lead to any obvious contradictions. It also is important to note that even the simple models discussed earlier may lead to considerable computational difficulties. The large numbers that can arise make it impossible to evaluate directly the sums in the given expressions, although the series themselves are clearly converging. It is, however, always possible and in most examples easy to simulate the sedimentation process and so obtain the relevant bed-thickness distributions. This ease of simulation makes it tempting to construct some complex models, but again the results of such simulations are indistinguishable from each other as well as from the simpler models. Clearly, the sedimentologic information contained in thickness measurements is limited, and perhaps at this stage it is not critical which sedimentation model is chosen.

An actual example will illustrate how the stratigrapher can benefit from studying a simple model in spite of all these difficulties. Bed-thickness measurements that were made in a vertical section and taken from a single locality (Sligo Limestone basin) are shown in Figure 8.5. A single bed from this section was followed laterally about 39 m and thickness measurements of this bed were taken 1 = m intervals (Figure 8.6).

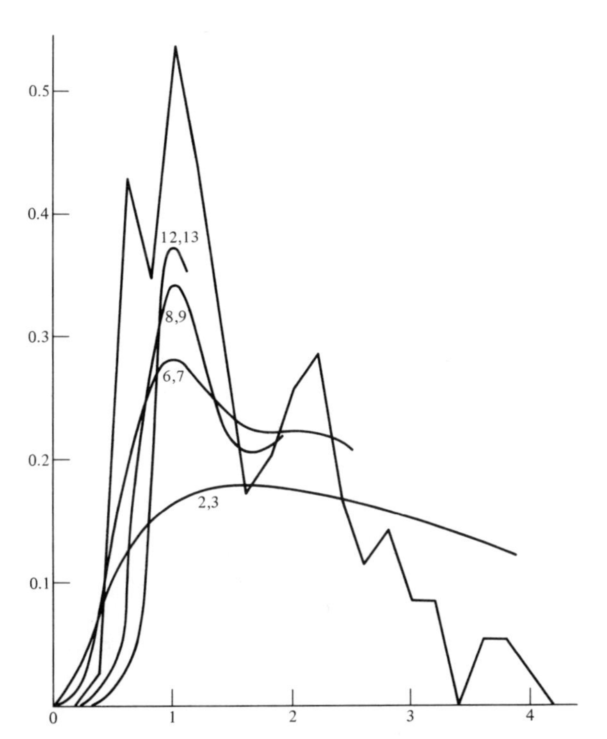


FIGURE 8.5 Measured bed thicknesses compared with calculated thickness distributions. Storm density = 2; steps are Gamma distributed with $\lambda$ = first figure, a = second figure.

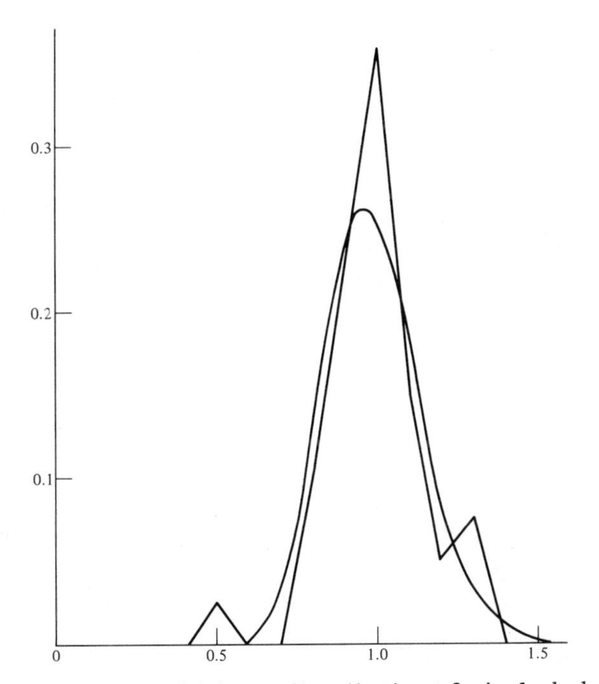


FIGURE 8.6 Thickness distribution of single bed.

All measurements are given in units of 5 cm so that the modes of both distributions are at 1.0.

A variable-step mechanism without recovery time is taken as the basic sedimentation model. It is known from petrographic studies that the number of steps in each bed in usually small, and a reasonable starting point is to assume that each bed represents, approximately, a constant time interval. Therefore, one may calculate bed-thickness distributions with low-storm densities under the assumption that the step thickness itself is Gamma distributed. Such results are shown in Figure 8.5; note that with increasing parameters of the step distribution, the computation becomes impossible because of limits in the computer's capacity for large numbers. Even if a complete fit of the observed distribution is not achieved, it can be seen how the model will behave with increasing further the parameters of the step distribution. Calculations were carried out with storm densities of 1, 2, 3, and $\rho = 2$ gave the nearest approximations. Using simulation methods, one can obtain a reasonably good fit by setting $\rho = 1$, $\lambda = 20$, and $a = 21$.

It is determined that the dispersion of the calculated curves that come closest to the observed modal frequency densities is smaller than in the measured distributions. If one examines the lateral variation of a single bed (which, according to theory, should provide an estimate of the step distribution), one notes an even smaller dispersion. In reality, fitting a Gamma distribution to the single-bed data gives the parameters $\lambda = 36.1$ and $a = 30$, which reflect the fact that the distribution is nearly symmetrical. If these values were introduced into the compound Poisson model, some sharp peaks would result. Accordingly, one may come to the conclusion that beds do not represent constant time intervals, as was initially assumed.

Thickness distributions closer to the observed ones can be obtained if the time taken for the formation of the bed is itself a random variable. Because direct calculation is beyond the computer's capacity, thickness distributions were again simulated. A variable-step model was used but the number of steps in each bed was determined by sampling from a negative binomial-distributed random array. The procedure corresponds to sampling with a variable counting interval. Preliminary results show that a standard deviation of 0.3 to 0.5 in the time-event distribution will lead to good approximations of the observed data (Figure 8.7).

The conclusions may be summarized as follows:

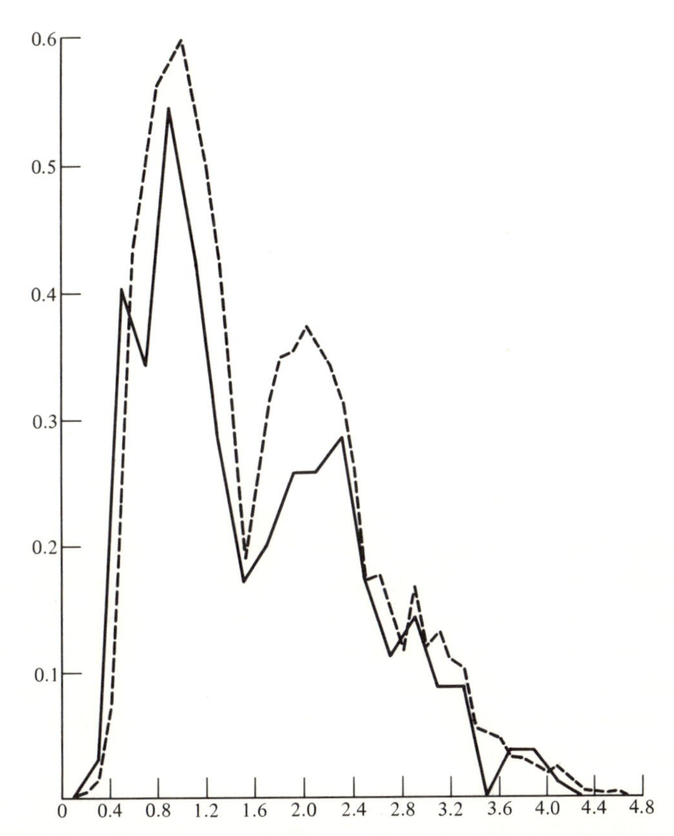


FIGURE 8.7 Measured bed thickness compared with simulated variable step model $\rho = 1$, $\lambda = 20$, $a = 21$. Time intervals are normally distributed with $\mu = 1.0$ and $\tau = 0.5$.

1. Each bed is the result of a small number of sedimentation steps.
2. The step thickness is approximately normally distributed with $\sigma = 0.152$ and $\mu = 1.0$.
3. The time taken for the formation of each bed has a mean length of one unit and a coefficient of variation of approximately 50%. Absolute age determinations could be based on counting beds in a section, but the given coefficient of variation indicates a low accuracy for such a procedure. Of course the magnitude of the time unit is unknown.

These conclusions are tentative because the model ignores a number of factors that are involved in the formation of beds. For example, it is well known that in this particular area successive bed-thickness measurements show a strong serial correlation, and in a similar manner it was determined that lateral thickness measurements are autocorrelated. This is partly due to local thickness variations that are produced by a certain waviness of the bedding planes and partly to regional trends that operate over wider areas. It was determined also that the lateral thickness variation does depend upon the variation in the bed that immediately underlies the measured bed. This indicates that compaction and diagenetic processes will contribute to the ultimate thickness of the bed and thus blur the picture of primary sedimentation. If these additional sources of variation exist, then the dispersion of the observed thickness distribution must be higher than predicted by theory. Therefore, there is less latitude for the possible variation of the time interval recorded by the bedding planes. In other words, the coefficient of variation given for the series of time events must be an upper limit.

Finally, one may consider the usefulness of the model in making stratigraphic correlations by the method of matching bed thicknesses. The lateral thickness distribution can be used in this problem for calculating the significance at which one can accept any match. According to the model used, a bed cannot disappear completely. Such a situation could be modeled by introducing a variable recovery time after each storm, which would make it possible that no sedimentation step takes place and therefore no bed would be recorded. Disappearance of a bed also could be caused by the incomplete recording of marker horizons in some localities. This is probably the situation in the examined limestones. Complications arise because it is likely that the process of a lithologic change equivalent to the production of a marker horizon may influence the sedimentation rates. Models involving two sedimentation states (alternating renewal processes) will be needed to deal with this situation.

## CONCLUSIONS

Only simple sedimentation models have been discussed, but an examination of the implications of such models can calrify questions of fundamental stratigraphic importance. It is interesting to note that the usually assumed situation of completely random time events between lithologic changes can be tested. Any further conclusions regarding time events will depend on how seriously the procedure of simplification has affected the performance of the model. No geologist will put too much reliance on conclusions similar to the ones reached in the previous paragraphs until it can be demonstrated that the model works in practice and that it is possible, for example, to predict with the model the vertical and lateral variations of bed thickness. One of the more important assets of the quantitative approach is that it indicates clearly where more field data are needed. In this particular study it was determined that more information on actual recent sedimentation processes is needed and that there are not enough quantitative data about the lateral behavior of sedimentary beds.

## REFERENCES

Bharucha-Reid, A. T., 1960, Elements of the theory of Markov processes and their application: McGraw-Hill Book Co., New York, 468 p.

Cox, D. R., 1962, Renewal theory: Methuen and Co., London, 142 p.

Feller, W., 1957, An introduction to probability theory and its applications, v. 1: John Wiley & Sons, New York, 461 p.

Jacod, J., and Joathon, P., 1972, Conditional simulation of sedimentary cycles in three dimensions, in Mathematical models of sedimentary processes: Plenum Press, New York, p. 139-165.

Schwarzacher, W., 1972, The semi-Markov process on a general sedimentation model, in Mathematical models of sedimentary processes: Plenum Press, New York, p. 247-268.

# Joint Distribution Functions for Link Lengths and Drainage Areas

J. S. Smart

ABSTRACT

The elementary link lengths $\ell$ and associated drainage areas a in drainage basins may be considered as correlated random variables drawn from a bivariate population. A theoretical model for the joint distribution is developed with the use of the assumption that $\ell$ and a depend, in turn, on other random variables, some of which they have in common. Some properties of the related quantities $\ell/a$ and $\ell^2/a$ also are derived, and some quantitative results that can be compared with observations on natural networks are obtained.

## INTRODUCTION

One of the goals of drainage-basin geomorphology is an understanding of the structure of channel networks. Here the term "structure" is used specifically to mean the two-dimensional topologic and geometric properties of channel networks and drainage basins. The basic elements from which drainage basins are constituted are the individual channel links and their associated drainage areas, as illustrated in Figure 9.1. A recent paper (Smart, 1972a) has suggested that the structure of channel networks can be characterized effectively by certain dimensionless parameters derived from measurements of the link lengths and drainage areas. Thus, for a set of N links there will be N pairs of measurements $(\ell_j, a_j)$, where the subscript j refers to the $j^{th}$ link.

It was pointed out that $\ell_j$ and $a_j$ may be regarded as correlated random variables drawn from a bivariate population. The purpose of this paper is to offer some considerations about the possible form of the joint distribution function. We also are interested in understanding the probability distributions of certain functions of the lengths and areas that are useful in the analysis of drainage-basin structure.

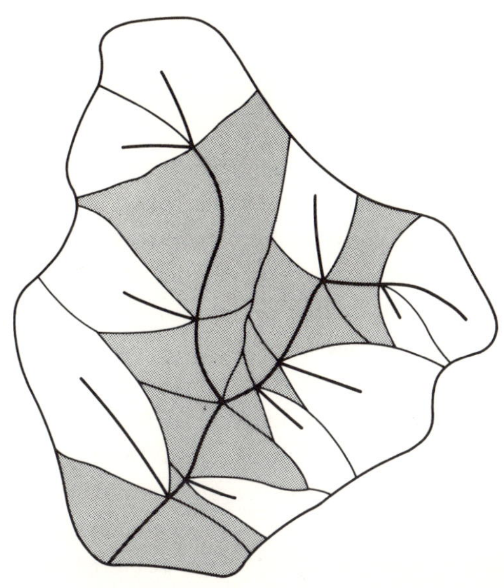

FIGURE 9.1 Channel network for Pritchard Hollow, Oleona, Pennsylvania, Quadrangle. Heavy lines indicate channels and basin boundary; light lines indicate internal drainage divides. White area drains into exterior links and solid areas into interior links.

The microscopic drainage density $\delta_j$ is given by

$$\delta_j = \ell_j / a_j \qquad j = 1, 2, 3, \ldots, N \tag{9.1}$$

An important related dimensionless parameter is

$$\phi_j = \ell_j^2 / a_j \tag{9.2}$$

The total channel length L and total drainage area A are, of course, given by

$$L = \sum_{j=1}^{N} \ell_j \tag{9.3}$$

and

$$A = \sum_{j=1}^{N} a_j \tag{9.4}$$

The macroscopic analogs of $\delta$ and $\phi$ are

$$D = L/A = \overline{\ell}_j / \overline{a}_j \tag{9.5}$$

and

$$K = L^2/NA = \overline{\ell_j^2} / \overline{a}_j \tag{9.6}$$

where the overbars indicate averages over the sample of N measurements.

## OBSERVATIONS ON NATURAL NETWORKS

It is always desirable that the construction of theoretical models be guided by the results of observations on actual systems. Unfortunately, although there exists a considerable literature on the quantitative properties of drainage networks, an examination of the available data does not give us many clues as to the possible nature of the joint distribution function for link lengths and areas.

One general feature of the data is that the marginal distributions of both exterior and interior link lengths are right skewed, but interior links generally have a higher proportion of short lengths. Results of previous measurements are reviewed in some detail by Shreve (1969, p. 400) and Smart (1972b, Sec. IIIA). The most comprehensive study is that of Krumbein and Shreve (1970, p. 82-91), who measured exterior and interior link lengths for 153 magnitude-5 networks and 30 magnitude-10 networks in eastern Kentucky. They determined that the data could be fairly well approximated by Gamma densities; maximum-likelihood estimates indicated shape factors near 2 for interior link lengths and appreciably larger (3.5 to 7.5) for exterior link lengths.

In the dimensionless parameter study (Smart, 1972a), measurements of exterior and interior link lengths were made on 10 networks ranging in magnitude from 130 to 321. These data were fitted to Gamma distributions by the maximum-likelihood method. In general, the shape-factor estimates were consistent with those of Krumbein and Shreve, and for two of the networks that came from the same region in Kentucky, the shape factors for interior links were 2.06 and 2.07 and 4.81 and 5.60 for exterior links.

One characteristic of exterior link lengths measured on 1:24,000 maps is that there is almost always a nonzero minimum value, typically around 100 to 150 ft, but sometimes appreciably larger. This result may be a real geomorphologic effect, or it may be due to faulty criteria for identifying first-order channels, or it may be a consequence of map-editing conventions, or it may be that all three factors are involved. In any event, the observation suggests that exterior link length distributions might be better approximated by Pearson III functions, i.e., Gamma distributions with an adjustable location parameter. The results of using the maximum-liklihood method to estimate three parameters (shape, scale, and location) were inconclusive. For five of the ten networks, location parameters in the range of 100 to 200 ft were estimated, shape parameters were brought into the range of 2 to 3, and the goodness of fit (as determined by $\chi^2$ tests) was considerably improved. For the remaining five networks, the maximum liklihood and best fits were obtained with the original two-parameter Gamma distributions.

There are essentially no data on the distributions of exterior and interior drainage areas. Maxwell (1960) gives the means and standard deviations of first-order drainage areas for five networks in the San Dimas Experimental Forest, California. If the areas are assumed to be Gamma distributed, moment-method estimates based on Maxwell's data yield shape factors between 1.1 and 1.8. Shreve (1969, p. 410-412) showed that the assumption that both exterior and interior areas are Gamma distributed with shape factor 2 (but with different means) is in good agreement with Schumm's observations (1956) on first- and second-order basin areas in the Perth Amboy network.

About the only conclusion that can be deduced from the observations is that the joint distribution function of $\ell$ and a should be right skewed in both margins. Because the Gamma densities are mathematically tractable and because they give reasonable fits to the link-length data, we shall look for joint distribution functions that are bivariate generalizations of the Gamma distribution.

Previous investigations of bivariate Gamma distributions have generally started from the observation that the sum of squares of n-independent standard normal variables is Gamma distributed with shape factor n/2, and proceeded by transformation on the well-known multivariate normal distribution (see, for example, Vere-Jones, 1967, and references cited there). This method is not satisfactory for our purposes, mainly because it has no rational connection with the problem of link lengths and areas, but also because the density functions derived in this manner are generally in the form of infinite series that are inconvenient for numerical calculations, even with the aid of a computer.

## THE DAVID-FIX MODEL

In trying to find a model more relevant to our problem, we may note that the variables $\ell$ and a, although undoubtedly highly correlated, do have different degrees of dependence on the various processes involved in the generation of drainage networks. As a simple example, in a stream network developing by headward growth and branching, a link length is determined by the locations of two successive junction (branching) points; the associated area, on the other hand, depends on the junction angles as well as the locations. In general, the random variables $\ell$ and a may be regarded as depending on other random variables, but not in exactly the same manner and not necessarily on exactly the same set of other variables. David and Fix (1961) have proposed, in another context, a correlated bivariate Gamma distribution that incorporates these general ideas in a simple, specific form. (A similar but slightly more restricted distribution was discussed by Cherian, 1941 .) The David-Fix model is described in some detail in the following sections and its predictions concerning link length-area relationships are investigated.

Let x, y, and z be independent random variables with the respective density functions $\gamma(x;P)$, $\gamma(y:Q)$, and $\gamma(z;R)$, where

$$\gamma(x;P) = \frac{x^{P-1}\, e^{-x}}{\Gamma(P)} \qquad x > 0, P > 0 \tag{9.7}$$

is the one-parameter Gamma probability density. We then may define a related set

$$u = x + y \qquad v = x + z \qquad w = x$$

The variables u and v are considered as dimensionless measures of $\ell$ and a, respectively. As x, y, and z are independent and as the Jacobian of the transformation is unity, the standard procedures for transforming variables (Hahn and Shapiro, 1967 , chap. V) give for the combined density function of u, v, and w

$$p(u,v,w;P,Q,R) = \frac{1}{\Gamma(P)\Gamma(Q)\Gamma(R)}\, w^{P-1}\, (u-w)^{Q-1}\, (v-w)^{R-1}\, e^{-(u+v-w)} \tag{9.8}$$

The bivariate function $p(u,v;\, P,Q,R)$ then can be obtained by integrating out the dependence on w. Because $w = x \leq u,v$, the integration must be performed differently in the two halves of the octant separated by the plane $u = v$.

$$p(u,v;P,Q,R) = \frac{e^{-(u+v)}}{\Gamma(P)\Gamma(Q)\Gamma(R)} \int_0^u w^{P-1}\, (u-w)^{Q-1}\, (v-w)^{R-1}\, e^{w} dw \tag{9.9A}$$

$$u < v$$

$$= \frac{e^{-(u+v)}}{\Gamma(P)\Gamma(Q)\Gamma(R)} \int_0^v w^{P-1}\, (u-w)^{Q-1}\, (v-w)^{R-1}\, e^{w} dw \tag{9.9B}$$

$$v < u$$

We can obtain the marginal density function $g(v;P,R)$ by defining the additional variable

$$t = x/(x+z) = x/v \qquad 0 < t < 1$$

transforming $\gamma(x;P)\gamma(z;R)$ into a function of u and t, and integrating with respect to t. The result is

$$g(v;P,R) = \gamma(v;P + R) \tag{9.10A}$$

By symmetry

$$f(u;P,Q) = \gamma(u;P + Q) \tag{9.10B}$$

Thus, the marginal distributions are Gamma densities, as desired. The conditional distributions are defined by

$$f(u|v;P,Q,R) = \frac{p(u,v;P,Q,R)}{\gamma(u;P + Q)} \tag{9.11A}$$

and

$$g(v|u;P,Q,R) = \frac{p(u,v;P,Q,R)}{\gamma(v;P + R)} \tag{9.11B}$$

These results can be used to show that the regression relations are linear

$$E\ u|v;P,Q,R = Q + P/(P + R)v \tag{9.12A}$$

$$E\ v|u;P,Q,R = R + P/(P + Q)u \tag{9.12B}$$

where $E[x]$ is the expected value of x. Also

$$\mathrm{cov}[u,v;P,Q,R] = P \tag{9.13}$$

from which it follows that the correlation coefficient is

$$\rho = \frac{P}{[(P + Q)\ (P + R)]^{1/2}} \tag{9.14}$$

Equations 9.9 to 9.14 contain the main features of the David-Fix model. At this point, it may be remarked that the two principal disadvantages of the model are the discontinuity in functional form indicated by Equations 9.9A and 9.9B and the fact that the correlation coefficient cannot be adjusted freely, but is completely determined by the values of P, Q, and R.

## PROPERTIES OF RELATED VARIABLES

For the link length-area distribution problem, we also are interested in the properties of

$$s = u/v = t + y/v$$

and

$$\phi' = u^2/v = t^2v + 2ty + y^2/v$$

The variable s is a dimensionless measure of the microscopic drainage density $\delta$ and $\phi'$ is proportional to $\phi$. The mean values of both variables can be calculated with the aid of the following well-known results.

If the variable x has the density function $\gamma(x;P)$, then

$$E(x;P) = P$$

$$E(x^2;P) = P(P + 1)$$

$$E(1/x;P) = 1/(P - 1) \qquad P > 1$$

The variable t is a beta variate of the first type with parameters P and R (Weatherburn, 1947, p. 153-154). Then

$$E(t;P,R) = P/(P + R)$$

$$E(t^2;P,R) = \frac{P(P + 1)}{(P + R)\ (P + R + 1)}$$

Thus

$$E(s;P,Q,R) = E(t;P,R) + E(y;Q)E(1/v;P + R)$$

because y and v are independent. Then

$$E(s;P,Q,R) = \frac{P}{P + R} + \frac{Q}{P + R - 1} \qquad P + R > 1 \tag{9.15}$$

Similarly

$$E(\phi';P,Q,R) = E(t^2v;P,R) + 2E(ty;P,Q,R) + E(y^2/v;P,Q,R)$$

The transformation described for obtaining Equation 9.10 shows that t and v are independent variables. Thus

$$E(\phi';P,Q,R) = E(t^2;P,R)E(v;P + R) - 2E(t;P,R)E(y;Q) + E(y^2;Q)E(1/v;P + R)$$

$$= \frac{P(P + 1)}{P + R + 1} + \frac{2PQ}{P + R} + \frac{Q(Q + 1)}{P + R - 1} \qquad P + R > 1 \tag{9.16}$$

The density function for s can be derived for certain specific situations by making the transformation $s = (x + y)/(x + z')v = x + z$, and $r = y/(x + z)$. Then

$$h(s,v,r;P,Q,R) = \frac{1}{\Gamma(P)\Gamma(Q)\Gamma(R)} v^{P+Q+R-1} (s - r)^{P-1} \times r^{Q-1} (1 - s + r)^{R-1} e^{-v(1+r)} \tag{9.17}$$

and integrating v out

$$h(s,r;P,Q,R) = \frac{\Gamma(P + Q + R)}{\Gamma(P)\Gamma(Q)\Gamma(R)} \frac{(s - r)^{P-1}\ r^{Q-1}\ (1 - s + r)^{R-1}}{(1 + r)^{P+Q+R}} \tag{9.18}$$

with

$$s - 1 < r < s \qquad s \geq 1$$

$$0 < r < s \qquad s < 1$$

Thus, the density function $h(s;P,Q,R)$ must be defined in two parts just as for u and v. We have not been able to perform a general integration over r, but explicit results can be obtained easily for small integer values of P, Q, and R, which are those of most interest.

## SOME NUMERICAL RESULTS

In order to apply the David-Fix model to real networks, some relations must be established between u, v, s, $\phi'$, and their respective counterparts $\ell$, a, $\delta$, and $\phi$.

Let $\ell = \ell_o u$ and $a = a_o v$, where $\ell_o$ and $a_o$ have the dimensions of length and area, respectively. Then

$$E(\ell;P,Q) = \ell_o E(u;P + Q) = (P + Q)\ell_o \tag{9.19A}$$

$$E(a;P,R) = a_o E(v;P + R) = (P + R)a_o \tag{9.19B}$$

If we use the sample means $\bar{\ell}$ and $\bar{a}$ as estimates of $E(\ell)$ and $E(a)$, then

$$\ell_o = \bar{\ell}/(P + Q) \qquad \text{and} \qquad a_o = \bar{a}/(P + R)$$

In terms of $\ell_o$ and $a_o$, the drainage density D is

$$D = \frac{\bar{\ell}}{\bar{a}} = \frac{(P + Q)\ell_o}{(P + R)a_o} \tag{9.20}$$

(It should be noted that D is a biased estimator of $E(\ell)/E(a)$ and that various procedures have been developed for correcting the bias; see, for example, Rao and Rao (1973) and references cited there.) The expected value of the microscopic drainage density is

$$E(\delta;P,Q,R) = E\left[\frac{\ell_o u}{a_o v};P,Q,R\right] = \frac{\ell_o}{a_o} E(s,P,Q,R)$$

$$= \left[\frac{P}{P + R} + \frac{Q}{P + R - 1}\right] \frac{\ell_o}{a_o} \qquad P + R > 1 \tag{9.21}$$

Again, if we accept $\bar{\delta}$ as an estimator of $E(\delta)$, we obtain a result that can be checked against observation

$$\frac{\bar{\delta}}{D} = 1 + \frac{Q}{(P + Q)\ (P + R - 1)} \qquad P + R > 1 \tag{9.22}$$

The review of length and area measurements in the introductory section suggests that, for interior links, $P + Q$ is approximately 2 and that a few indirect observations are consistent with $P + R \quad 2$ also. In order to get some quantitative feeling for the predictions of the David-Fix model, we shall assume that $P = Q = R = 1$. For Equation 9.22, this gives $\overline{\delta}/D = 3/2$. As a comparison with natural networks, observations by Maxwell (1960) on first-order streams in five basins in the San Dimas Experimental Forest, California, give values of $\overline{\delta}/D$ ranging between 1.21 and 1.28.

We also have

$$K = \frac{\overline{\ell}^2}{\overline{a}} = \frac{(P + Q)^2 \ell_o^{\,2}}{(P + R) a_o} \tag{9.23}$$

whereas

$$E(\phi;P,Q,R) = E\left[\frac{\ell_o^2 u^2}{a_o v}; P,Q,R\right] = \frac{\ell_o^2}{a_o} E\,(\phi';P,Q,R)$$

$$= \left[\frac{P(P + 1)}{P + R + 1} + \frac{2PQ}{P + R} + \frac{Q(Q + 1)}{P + R - 1}\right] \frac{\ell_o^2}{a_o} \qquad P + R > 1 \tag{9.24}$$

The expression corresponding to Equation 9.22 for $\overline{\phi}/K$ cannot be put into simple algebraic form. For $P = Q = R = 1$, however, we have $\overline{\phi}/K = 11/6$. From Equation 9.9

$$p(u,v;1,1,1) = e^{-v}(1 - e^{-u}) \qquad u < v \tag{9.25A}$$

$$= e^{-u}(1 - e^{-v}) \qquad u > v \tag{9.25B}$$

This density function, which is symmetric in u and v, is plotted in Figure 9.2. The positive correlation between u and v is clearly indicated by the shape of the contour lines. The corresponding joint density function for $\ell$ and a will be symmetric about the line

$$\ell = (P + R)/(P + Q)Da \tag{9.26}$$

We can obtain the density function $h(s;1,1,1)$ by substituting $P = Q = R = 1$ in Equation 9.18 and integrating r between the limits indicated.

$$h(s;1,1,1) = 1 - \frac{1}{(1 + s)^2} \qquad s < 1 \tag{9.27A}$$

$$= \frac{1}{s^2} - \frac{1}{(1 + s)^2} \qquad s \geq 1 \tag{9.27B}$$

This function, which is plotted in Figure 9.3, rises rapidly from zero,

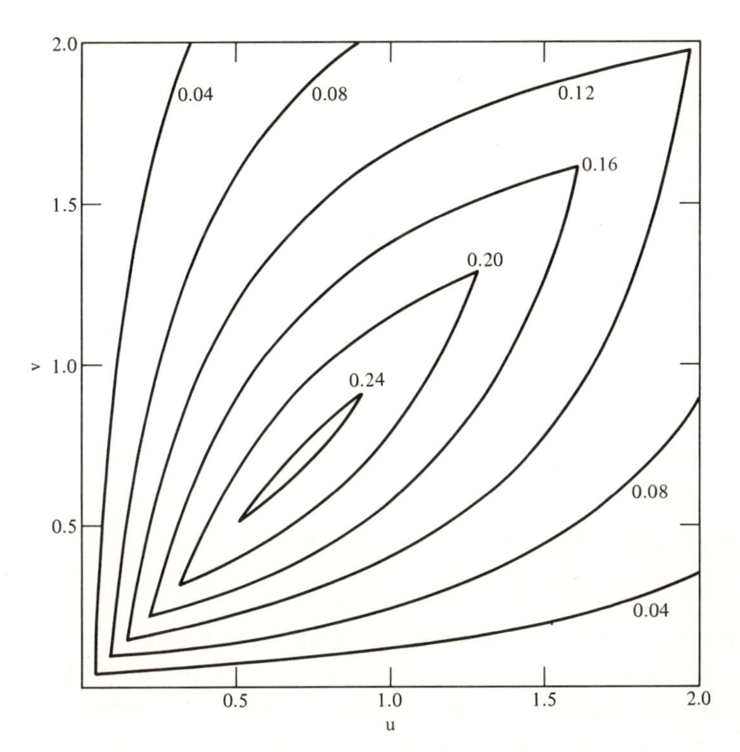


FIGURE 9.2 Contours of joint density function p(u,v;1,1,1).

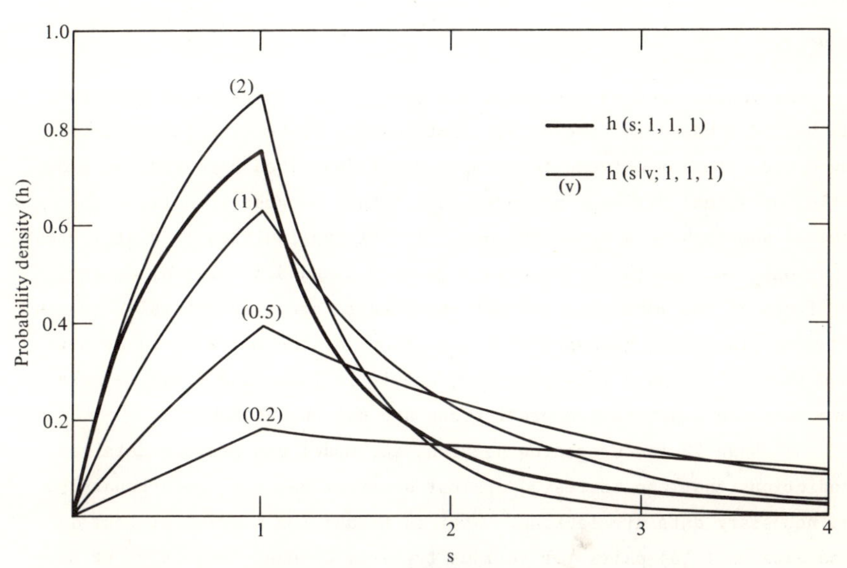


FIGURE 9.3 Heavy line = density function h(s;1,1,1). Light lines = conditional density functions h(s|v;1,1,1) for v = 0.2, 0.5, 1, and 2.

has a maximum at $s = 1$, and then decreases slowly with increasing s. Translating these results into terms of drainage densities, we have $\overline{\delta} = 3D/2$, whereas the most probable value is $\delta = D$. $E(s^2)$ diverges and thus s does not have a finite variance; in general, var(s) is not defined unless $P + R > 2$.

We also can obtain the conditional density function $h(s|v;1,1,1)$, which corresponds to the conditional probability density of $\delta$ for given a.

$$h(s|v;1,1,1) = \frac{p(s,v;1,1,1)}{g(v;2)} = \int \frac{p(s,v,r;1,1,1)dr}{\gamma(v;2)} \tag{9.28}$$

Using Equations 9.7 and 9.17, and performing the indicated integration over both sets of limits for r, we obtain

$$h(s|v;1,1,1) = 1 - e^{-vs} \qquad s < 1 \tag{9.29A}$$

$$= e^{-vs}\,(e^{v} - 1) \qquad s \geq 1 \tag{9.29B}$$

This function is shown in Figure 9.3 for several different values of v. The most interesting feature is the increase in relative probability of large s with decreasing v. This result is in qualitative agreement with observations on natural networks, where it is known that, for a given region, small elementary drainage areas tend to have large drainage densities.

## COMMENTS

The application of the David-Fix model to drainage-basin structure has been discussed in some detail, not because we believe that there is any close relation between the simple model described here and the properties of actual drainage networks, but rather because we believe the general approach is a promising one. The discontinuities in distribution functions, such as those shown in Figures 9.2 and 9.3, are, of course, artifacts of the model and are not expected to occur in natural networks. It seems plausible, however, that variations of the model could be developed that do not have this property, but it is important first to make sure that the quantitative predictions are not unreasonable.

As shown in the preceding section, the model can provide detailed predictions that can be tested against observation, but, unfortunately, the necessary data are lacking. What is needed for a critical test is good data on $(\ell,a)$ pairs for various types of channel networks. We have assumed for the purposes of discussion here that $\ell$ is marginally distributed as $\gamma(\ell;2)$ and a is marginally distributed as $\gamma(\ell;2)$, but the first assumption is not really justified and the second is hardly more than speculation. Our understanding of the problem will not progress much farther without better data and without rational analysis to limit the many possible empirical choices in the selection of a model.

## ACKNOWLEDGMENT

This research was supported in part by the Geography Programs, Office of Naval Research, ONR Contract No. N00014-70-C-0188, Task No. NR389-155.

## REFERENCES

Cherian, K. C., 1941, A bi-variate correlated gamma-type distribution function: Jour. Indian Math. Soc. (ns), v. 5, p. 133-144.

David, F. N., and Fix, E., 1961, Rank correlation and regression in a non-normal surface, in Proc. of the Fourth Berkeley Sym. on Mathematical Statistics and Probability, v. 1: Univ. California Press, Berkeley, p. 177-197.

Hahn, G. J., and Shapiro, S. S., 1967, Statistical models in engineering: John Wiley & Sons, New York, 355 p.

Krumbein, W. C., and Shreve, R. L., 1970, Some statistical properties of dentritic channel networks: Tech. Rep. 13, ONR Task No. 389-150, Dept. Geol. Sciences, Northwestern Univ. and Sp. Proj. Rept., NSF Grant 6A-1137, Dept. Geology, UCLA, 117 p.

Maxwell, J. C., 1960, Quantitative geomorphology of the San Dimas National Forest, California: Proj. NR389-042 Tech. Rept. 19, Dept. Geology, Columbia Univ., 95 p.

Rao, P. S. R. S., and Rao, J. N. K., 1973, Small sample results for ratio estimators: Biometrika, v. 58, no. 3, p. 625-630.

Schumm, S. A., 1956, Evolution of drainage systems and slopes in badlands at Perth Amboy, New Jersey: Geol. Soc. America Bull., v. 67, no. 5, p. 597-646.

Shreve, R. L., 1969, Stream lengths and basin areas in topologically random channel networks: Jour. Geology, v. 77, no. 4, p. 397-414.

Smart, J. S., 1972a, Quantitative characterization of channel network structure: Water Resources Res., v. 8, no. 6, p. 1487-1496.

Smart, J. S., 1972b, Channel networks, in Advances in hydroscience, v. 8: Academic Press, New York, 359 p.

Vere-Jones, D., 1967, The infinite divisibility of a bivariate Gamma distribution: Sankhya, Ser. A, v. 29, no. 4, p. 421-422.

Weatherburn, C. E., 1947, A first course in mathematical statistics: Cambridge Univ. Press, London, 271 p.

# 10 Applications of Random Process Models to the Description of Spatial Distributions of Qualitative Geologic Variables

P. Switzer

ABSTRACT

Complex spatial geologic patterns may be regarded as realizations of random processes. The estimated parameters of such processes serve as convenient summary characteristics of the observed geologic patterns, and provide a basis for their classification and comparison. The statistical properties of the estimates of process parameters, e.g., prevalence and patchiness, are related to the rate and method of sampling, as well as to the model of the process itself. The paper contains an investigation of these relationships. The scale of phenomena to which these investigations apply range from the texture of rock thin sections to satellite surface imagery.

## INTRODUCTION

The use of random process models to describe spatial variation is not new to earth scientists. In particular, we note the work of Matheron (1965), Watson (1971), and Dacey (1964) and ecologists Matern (1960) and Pielou (1964), among others. In this paper we deal only with models of qualitative spatial variation, i.e., variation representable by simple two-color or multicolor patterns.

In particular, we will be interested in estimating properties of such patterns from discrete sample-point data, summarizing the earlier work of Matern (1960) and Switzer (1967), as well as some recent contributions. Some important applications where the data are characteristically obtained at discrete locations are soil samples, grab rock samples, points on a thin section, weather stations, digitized photographs and other imagery, and scoop samples from the ocean bottom.

It is convenient to regard the underlying qualitative variation as a multicolor pattern or, where appropriate, as a two-color black-white pattern. The discretely spaced sample data might be used to (1) estimate the overall area proportions of various colors as in thin-section modal analyses; (2) obtain an estimated reconstruction of the actual pattern of variation; (3) obtain an estimated measure of overall pattern complexity, especially when comparing several patterns on the basis of complexity; or (4) obtain estimates of parameters of the pattern-generating process for their own sake. Each of these uses of the sample data involves the estimation of unknown quantities. We shall examine how such estimates may be constructed and how the statistical properties of the estimates are related to the underlying variation and the density of sampling.

## ESTIMATING AREAL PROPORTIONS

Let $p_j$ = areal proportion of a specified color j (say black)

$n_j$ = number of data points counted of the specified color j

N = total number of data points

Then the usual estimator of $p_j$ is $\hat{p}_j = n_j/N$. If the data-point locations are randomly selected, then an approximate 90% confidence interval for p is usually given by

$$\hat{p}_j \pm 1.65\ \text{S.E.}(\hat{p}_j) \tag{10.1}$$

where S.E.$(\hat{p}_j)$ is itself estimated by $\sqrt{\hat{p}_j(1 - \hat{p}_j)}/\sqrt{N}$, provided $\hat{p}_j$ is not too close to 0 or 1.

This confidence interval does not depend on the actual configuration of the data points obtained, but rather it is a statement about the average performance of the estimate under repeated random samplings of N data points.

However, it would seem more useful to be able to construct confidence intervals appropriate to the actual data-point configuration, especially because data locations are not usually selected at random. Indeed, as it usually turns out, a regularly spaced grid of points will allow one to use a shorter confidence interval for a given N than the interval given by Equation 10.1. In this situation randomness is imparted either by allowing for a random displacement and orientation of the grid or, equivalently, by imparting suitable properties to the pattern-generating model. Specifically, let

$$p_{jj}(d) = \text{probability that two randomly selected points, distance d apart, are both color j} \tag{10.2}$$

Then, it can be shown (Matern, 1960) that the length of a confidence interval for $p_j$ can be expressed completely in terms of $p_{jj}(d)$ for a given sample-point configuration.

For d close to zero, $p_{jj}(d)$ will usually be nearly $p_j$, whereas for a large d, the two points will usually be independent so that $p_{jj}(d)$ will be nearly $p_j^2$. It is customary to model $p_{jj}(d)$ by a function that decreases steadily from $p_j$ to the value $p_j^2$. For example, we might take

$$p_{jj}(d) = p_j e^{-cd} + p_j^2(1 - e^{-cd}) \tag{10.3}$$

where c is a rate parameter. The correct confidence interval for $p_j$ depends on knowing the true form of $p_{jj}(d)$. Fortunately, it is not sensitive to $p_{jj}(d)$ as long as there is some flexibility of parametrization, as with the constant c in model expressed by Equation 10.3.

Thus, for N sample points arranged on a square grid with spacing s, it can be shown, by making suitable approximations in the general form as given by Matern (1960), for example, that a confidence interval estimator for $p_j$ is

$$\hat{p}_j \pm 1.65 \text{ S.E.}(\hat{p}_j) \tag{10.4}$$

where S.E.$(\hat{p}_j)$ is itself estimated by $0.47\sqrt{\hat{c}s}\sqrt{\hat{p}_j(1 - \hat{p}_j)}/\sqrt{N}$ and $\hat{c}$ is an estimate of the rate parameter c. It is important to note that the estimate of S.E.$(\hat{p}_j)$ in Equation 10.4 is appropriate only if $\hat{c}s$ is less than about 2, i.e., if there is an appreciable amount of dependence between adjacent sample points.

The problem now is to obtain a rate estimator $\hat{c}$ from the data. This can be accomplished indirectly by estimating $p_{jj}(d)$ for d = s; i.e., by estimating the probability that an adjacent pair of data points are both black. Thus

$$\hat{p}_{jj}(s) = \frac{\text{Number of adjacent sample pairs observed to be both color j}}{\text{Total number of adjacent sample pairs}} \tag{10.5}$$

Thus, solving for $\hat{c}s$ using Equation 10.3 and the estimated value of $\hat{p}_{jj}(s)$ and $\hat{p}_j$, we obtain

$$\hat{c}s = \log_e(\hat{p}_j - \hat{p}_j^2)/(\hat{p}_{jj}(s) - \hat{p}_j^2) \tag{10.6}$$

which, in turn, can be used in Equation 10.4 to obtain the approximate confidence interval.

For example, Figure 10.1A shows the areal distribution of two ages of rocks in a portion of northern New Mexico. This areal distribution was

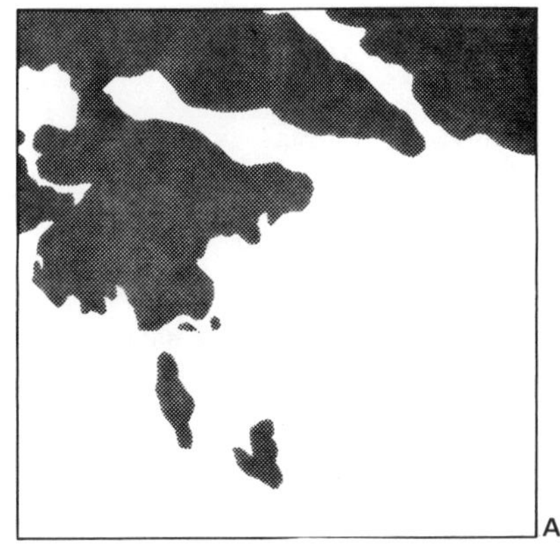


| | | | | | | | | | |
|---|---|---|---|---|---|---|---|---|---|
| B | B | B | B | B | B | W | B | B | B |
| W | B | W | B | B | B | B | W | B | B |
| B | B | B | B | W | W | W | B | W | B |
| B | B | B | B | B | B | W | W | W | W |
| B | B | B | B | W | W | W | W | W | W |
| W | B | B | B | W | W | W | W | W | W |
| W | W | W | W | W | W | W | W | W | W |
| W | W | W | W | W | W | W | W | W | W |
| W | W | W | W | W | W | W | W | W | W |
| W | W | W | W | W | W | W | W | W | W |

B

Precambrian undivided

Tertiary to Recent sediments and alluvium

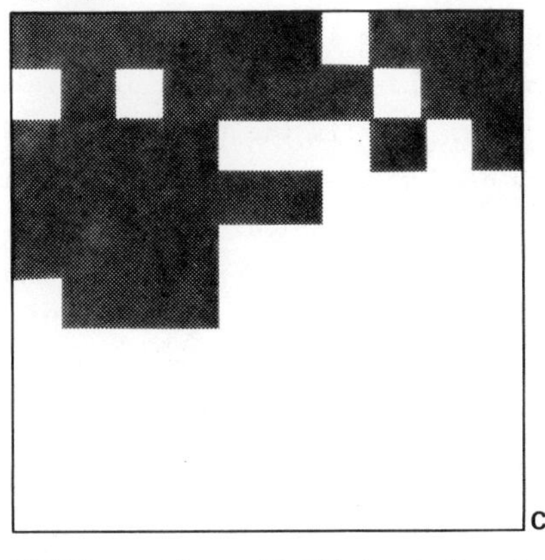


Precambrian undivided

Tertiary to Recent sediments and alluvium

FIGURE 10.1 A. Areal distribution geologic map of portion of northern New Mexico. B. Data obtained by sampling pattern of Figure 10.1A by square grid of 100 data points (B = observed black point Precambrian ; W = observed white point Tertiary to Recent ). C. Estimated pattern using nearest-point estimation with data of Figure 10.1B.

sampled as shown in Figure 10.1B by a square grid with spacing $s = 0.5$ in the scale of the map. There are altogether $N = 100$ sample points, and they give estimates $\hat{p}_b = 35/100$, $\hat{p}_{bb}(s) = 44/180$, $\hat{c}s = 0.63$, with a 90% confidence interval for $p_b$ given by $0.35 \pm 0.03$, where $p_b$ is the black area proportion.

It is interesting to note that not only is the usual form (shown in Equation 10.1) inappropriate if there is dependence between sample points, and not only are the confidence intervals given by 10.1 too wide, but also that an increase in the density of the sampling grid gives a proportionately larger increase in precision under Equation 10.4. Specifically, because the spacing s is itself porportional to $N^{-\frac{1}{2}}$, the width of the confidence interval under Equation 10.4 decreases as $N^{-3/4}$, rather than as $N^{-\frac{1}{2}}$, as in 10.1. For example, doubling the total number of sampling points in the grid should reduce the confidence interval length by about 40%, whereas using 10.1 would imply a length reduction of only 29%.

If the sampling grid is sparse relative to the pattern of spatial variation, then there generally will be little dependence between adjacent points. This will be indicated by estimates $\hat{p}_{jj}(s)$ and $\hat{p}_j^2$, which are close to each other, giving a large value for $\hat{c}s$ in Equation 10.6. In such situations, even though the sampling is on a grid, it is effectively similar to random sampling and the confidence intervals expressed in 10.1 then would be appropriate. Such is the typical situation when point-counting thin sections of fine-grain rock, where the frequency of two points falling on the same grain is low.

A different approach to obtaining confidence intervals for the proportion $p_j$ has been suggested by Demirmen (1969). His idea is to divide the sampled area into k subareas of equal size and then obtain a separate estimator $\hat{p}_{ji}$ in each subarea i = 1, 2, . . ., k. Reasoning that there will be little statistical dependence between the subarea estimators, even in the presence of point-to-point dependence, he forms 90% confidence intervals of the form

$$\hat{p}_j \pm 1.65 \text{ S.E.}(\hat{p}_j) \tag{10.7}$$

where S.E.$(\hat{p}_j)$ is estimated by $\sqrt{\Sigma_i(\hat{p}_{ji} - \hat{p}_j)^2/k(k-1)}$ and $\hat{p}_j$ is the usual overall sample proportion of black points. This method will be most successful with k > 5, say, and for patterns with substantial variation even within subareas. Demirmen applied this method with k = 6 to point-counting large-grain rocks.

## ESTIMATING PATTERN COMPLEXITY

It is not simple to designate an index of pattern complexity that will be suited to all applications. By complexity, we usually have in mind the spatial scale of variation. Thus, a pattern that has all its black area in one contiguous lump is less complex than another pattern with the same proportion of black area distributed in many scattered little lumps. This notion of complexity as a scale measurement also may be viewed as a measure of patchiness of the pattern.

One index that seems to incorporate these intuitive notions of complexity would be

$$x = \frac{\text{Total length of boundaries between different colors}}{(\text{Area of region})^{\frac{1}{2}}} \qquad (10.8)$$

The larger the value of x, the more complex the pattern is, and it is invariant to the choice of measurement unit. Of course, we will need some convention on how to measure boundary length, which we must assume is finite.

It can be shown, using the results of Matern (1960), that if the pattern is regarded as a realization of a random process, then

$$\text{mean of } x = \tfrac{1}{2}\pi \ (\text{area of region})^{\frac{1}{2}} Q'(0) \qquad (10.9)$$

where $Q'(0)$ = derivative of $Q(d)$ at $d = 0$

$Q(d)$ = probability that two points distance d apart are of different colors $= 1 - \Sigma p_{jj}(d)$

(It also can be shown that the function $Q(d)$, which measures the overall dependence in the pattern, necessarily increases concavely from zero. For large distances d, the dependence should be negligible and $Q(d)$ should therefore be nearly $\Sigma p_j(1 - p_j)$, as implied by independence.)

Because we are concerned here about estimation of pattern properties from discretely spaced data, we may wish to know how the complexity parameter x might be estimated from, say, a square data grid. Basically, this means estimating the derivative $Q'(0)$. If the model expressed by Equation 10.3 is assumed for the dependence functions $p_{jj}(d)$, then

$$\begin{aligned} Q(d) &= (1 - e^{-cd})\Sigma p_j(1 - p_j) \\ Q'(0) &= c\Sigma p_j(1 - p_j) \end{aligned} \qquad (10.10)$$

Using this $Q'(0)$ model, we can estimate all the proportions $p_j$ in the usual manner, and an estimator of c could be obtained as in Equations 10.5 and 10.6. However, this estimator of the rate c is particular to one color; the corresponding overall estimator of c for sample spacing s would be

$$\hat{c}s = \log_e \hat{T} - \log_e[\hat{T} - \hat{Q}(s)] \qquad \text{with}$$

$$\hat{T} = \Sigma\hat{p}_j(1 - \hat{p}_j) \qquad \text{and} \qquad (10.11)$$

$\hat{Q}(s)$ = proportion of pairs of adjacent sample points observed to be of different colors

Hence, the data-based estimator of total pattern complexity can be expressed as

$$\hat{x} = \tfrac{1}{2}\pi \times (\text{area of region})^{\frac{1}{2}}\ \hat{T},\hat{c} \tag{10.12}$$

with $\hat{T},\hat{c}$ as given in Equation 10.11. For two-color patterns, the estimator of $\hat{c}$ given in Equation 10.11 is equivalent to the estimator given in Equation 10.6 for either of the two colors.

Note that the estimator $\hat{Q}(s)$ will usually decrease as the sample spacing s decreases. For relatively dense grids, the estimator Equation 10.11 therefore can be approximated by

$$\begin{aligned} \hat{c}s &\doteq \hat{Q}(s)/\hat{T} \qquad \text{from which} \\ \hat{x} &\doteq \tfrac{1}{2}\pi \times (\text{area of region})^{\frac{1}{2}}\ \hat{Q}(s)/s \end{aligned} \tag{10.13}$$

Thus, in the context of the model expressed by Equation 10.3, which relates distance with dependence, it is seen that the confidence interval (Equation 10.4) for an estimated proportion $p_j$ is directly related to $\hat{x}$, the estimated complexity index, at least for sufficiently small grid spacing. That is, the length of the confidence interval is proportional to $\sqrt{\hat{x}}$. However, whereas the interval length is not particularly sensitive to a choice of model for $p_{jj}(d)$, the estimator of the complexity is not likewise insensitive. Nevertheless, one can use Equation 10.12 as an estimator of complexity in any event, but it will not, in general, be a good estimator of boundary length for arbitrary dependence models. If various different sample spacings s all gave approximately the same estimates $\hat{x}$, then presumably one would be satisfied with the measure.

## RECONSTRUCTING THE PATTERN

If discrete sample data are used in an attempted reconstruction of the actual underlying pattern, one obtains what might be called an estimated pattern. The difference between the estimated and underlying patterns might reasonably be measured by the index

$$L = \frac{\text{Total mismatched area}}{\text{Total area}} \tag{10.14}$$

where the mismatched area is that where the color of the estimated pattern does not coincide with the color of the underlying pattern. Of course, we cannot compute L unless we know the underlying pattern, and, even then, as in digitized photo- or map-sampling applications, it would be arduous. Remarkably enough, however, we can usually obtain a good estimate of L solely from the discretely spaced sample data.

The estimation of L is necessarily related to the method one uses to obtain the estimated pattern from the discrete data. For example, the estimated pattern may be constructed by assigning to each point in the region the color of the nearest observed data point. This method, called "nearest point," is simple and objective but it generates sharp-cornered boundaries between colors. This is not a great drawback when the sampling is dense, but even with light to moderate sampling it is not obvious that smoothly drawn freehand boundaries will necessarily give a more accurate estimated pattern accoding to the L criterion. Figure 10.1C shows an estimated pattern constructed by the nearest-point method, using the data of Figure 10.1B.

For nearest-point pattern estimation, it can be shown (Switzer, 1967) that the mean of L depends on the underlying pattern only through the dependence function Q(d), as defined in Equation 10.9. For example, with data points on a square grid with spacing s, we obtain

$$\text{mean of } L = \int_{-\frac{1}{2}}^{\frac{1}{2}} \int_{-\frac{1}{2}}^{\frac{1}{2}} Q(s\sqrt{u^2 + v^2})\, du dv \qquad (10.15)$$

It is important to note that we need to specify Q(d) only in the range $0 < d < s/\sqrt{2}$ to evaluate the integral in Equation 10.15. However, we should note further that the discretely spaced data allow us to estimate the dependence function Q(d) directly only for d = s = sample spacing (as in Equation 10.11), or even larger d values. Hence, the specification of a model for the shape of Q(d) for small d values would seem to be essential. Fortunately, the operation of integration means that expression 10.15 will not be particularly sensitive to a precise model specification, so long as our Q(d) is increasing concavely from Q(d) = 0 at d = 0 to $\hat{Q}(s)$ at d = s.

If we use the dependence model expressed by Equation 10.10, for example, then expression 10.15 for the mismatched proportion specializes and can be estimated as follows:

$$\hat{L} = \hat{I}\cdot\hat{T} \qquad \text{with}$$

$$\hat{I} = 1 - \int_{-\frac{1}{2}}^{\frac{1}{2}} \int_{-\frac{1}{2}}^{\frac{1}{2}} \exp\{-\hat{c}s\sqrt{u^2 + v^2}\}\, du dv \qquad (10.16)$$

with the estimators $\hat{c}s$ and $\hat{T} = \Sigma\hat{p}_j(1 - \hat{p}_j)$, as given in Equation 10.10. The integral $\hat{I}$ can be evaluated numerically and a short example, taken from Switzer (1967), is given in Table 10.1.

Applying the nearest-point procedure to the pattern and data of Figure 10.1B, we obtain the estimated pattern also shown in Figure 10.1C. The discrete data themselves gave estimates of $\hat{p}_b = 35/100$ and $\hat{p}_w = 65/100$ for black and white, respectively, giving $\hat{T} = 0.455$. The data estimate of

TABLE 10.1 Numerical evaluation of integral $\hat{I}$[a,b]

| $\hat{c}s$ | 0.3 | 0.4 | 0.6 | 0.8 | 1.0 | 1.4 | 2.0 |
|---|---|---|---|---|---|---|---|
| $\hat{I}$ | 0.11 | 0.14 | 0.20 | 0.26 | 0.31 | 0.40 | 0.52 |

[a] Refers to Equation 10.16.
[b] From Switzer, 1967.

$\hat{c}s$ was 0.63, so from Table 10.1, we obtain $\hat{I} \doteq 0.21$. Hence the estimate of the proportion of mismatched area is $\hat{L} = 9.6\%$. An actual planimetric calculation of the mismatched area of the two superimposed patterns gave $L = 9.2\%$. But it is important to note that $\hat{L}$ is obtained only from the sample data and without further knowledge of the actual pattern. We can infer also from Table 10.1 that if the number of sample points were quadrupled, say (reducing the sample spacing s by one-half), then the mismatched proportion would decrease from about 9.6% to about 5.1%.

So we see that the accuracy of an estimated map, in this context, depends only on the proportions of the various colors and the rate at which spatial dependence decreases - as was the situation also with confidence intervals for estimated proportions and with the pattern complexity measure. From the point of view of estimation in any of these three problems, the important remaining statistical question is the stability of the rate estimator $\hat{c}s$ in Equations 10.6 or 10.11.

There is not much that can be said from a theoretical point of view about the $\hat{c}$ estimator, except that its statistical properties will depend in a complicated manner on probabilities involving configurations of four points. We note that the principal source of variability in Equation 10.6 lies in the estimator $\hat{Q}(s)$, the proportion of different colored adjacent sample points; relatively, the estimator $\hat{T}$ will be stable. From purely geometric considerations, it can be deduced that the range of variation of $\hat{Q}(s)$ will not exceed $(\sqrt{2} - 1)100\%$ if the spacing s is small. In practice, however, the variation will usually be smaller. A general rule of thumb for statistical stability might be to increase the sampling density s unless the estimated value $\hat{c}s$, as given in Equation 10.11, is already less than about 2.

It is interesting to compare the length of the boundary between colors in the estimated map with the estimated boundary length as given by the complexity estimator (Equation 10.12). These two quantities are, respectively

$$2\hat{Q}(s) \times (\text{area of region})/s$$

and

$1/2\pi$ x (area of region)$\hat{T},\hat{c}$

As noted in Equation 10.13, for dense sampling, $\hat{T},\hat{c}$ is approximated by $\hat{Q}(s)/s$. Hence, in this situation, the boundary length in the estimated pattern consistently overestimates the true boundary length by a factor $4/\pi$. This overestimation is created by the sharp angles in nearest-point pattern reconstruction with samples on a square grid. However, the area proportions of the various colors in the estimated pattern will correspond to the usual proportion estimates $\hat{p}_j$.

## OTHER TOPICS

This paper has been concerned with spatial models for qualitatively changing phenomena mainly through the role played by such dependence functions as Q(d) in Equation 10.9, which are simply related to the spatial correlation function. A further discussion of this and other properties of spatial models may be found in Matern (1960) and Switzer (1965, 1967), for example. The general question of whether a particular model provides an appropriate description of a given geologic pattern can only be answered relative to the type of description one has in mind. Oçcasionally, the physical situation is sufficiently circumscribed to make the model of spatial variation strongly specified, as, for example, in the paper by Gilbert (1962) on crystal packing.

Also, we have been concerned here almost entirely with problems arising from estimation derived from discretely spaced sample points, especially points on a square grid. General configurations of sample points introduce more complexity in estimation problems, but a complexity that is not unmanageable. As a practical matter, it may be more convenient to sample on an elongated rectangular grid when sampling along transects in the field, from a ship, or along the flight path of an aircraft or satellite. In such situations the spacing between sample data along one direction (along transects) will be closer than the spacing in the perpendicular direction (between transacts). In general, for a given total number of sample points, there is a loss of efficiency in estimation as the grid becomes more elongated.

For example, if the ratio of grid spacings is 10:1, then we would need about four times as many sample points as are needed with a square grid to achieve a given precision of estimation (see Switzer, 1967 for additional details and tables). However, the travel economies of transect sampling may make it cheaper to obtain these four times as many data points. It also can be shown that randomly located data points have about the same estimating efficiency as a 3:1 rectangular sampling grid. The gain attributed to triangular sampling grids compared with square grids is negligible.

In conclusion, we refer to pattern-estimation problems where the data themselves are subject to error, i.e., if there are nonnegligible probabilities of an incorrect "color" assignment even at the data points. Such is the situation in remote-sensing surface mapping where "colors" are assigned on the basis of fairly noisy spectra; it also might occur because of location errors, for example, in ocean-bottom sampling. In these situations we have more than just an interpolation problem. An overview of the effect this has on estimation can be found in Switzer (1971) and a detailed inquiry into alternation along one dimension in this context is found in Devore (1971).

## REFERENCES

Dacey, M. F., 1964, Measures of contiguity for two-color maps: Technical Rept., Northwestern Univ., 25 p.

Demirmen, F., 1969, Petrographic and statistical study of part of Pennsylvanian Honaker Trail Formation: unpubl. doctoral dissertation, Stanford Univ., 531 p.

Devore, J., 1971, Noisy Markov chains: unpubl. doctoral dissertation, Stanford Univ., 85 p.

Gilbert, E. N., 1962, Random subdivisions of space into crystals: Ann. Math. Stat., v. 33, p. 958-972.

Matern, B., 1960, Spatial variation: Medd. Statens Skogsforskningsinstitut, 144 p.

Matheron, G., 1965, Les variables regionalises et leur estimation: Masson & Cie., Paris, 306 p.

Pielou, E. C., 1964, The spatial pattern of two-phase patchworks of vegetation: Biometrics, v. 20, p. 156-167.

Switzer, P., 1965, A random set process with a Markovian property: Ann. Math. Stat., v. 36, p. 1859-1863.

Switzer, P., 1967, Reconstructing patterns from sample data: Ann. Math. Stat., v. 38, p. 138-154.

Switzer, P., 1971, Mapping a geographically correlated environment: Office Naval Res. (NR-342-022), Tech. Rept. No. 145, 38 p.

Watson, G., 1971, Trend-surface analysis: Jour. Math. Geology, v. 3, no. 3, p. 215-226.

# Markov Models of Repose-Period Patterns of Volcanoes

# 11

F. E. Wickman

ABSTRACT

Six renewal type Markov models illustrating different idealized patterns of volcanic activity are examined: (1) simple two-stage volcano without persistent activity (Fuji type), (2) volcano with "loading time" and dormancies (Hekla type), (3) volcano with "loading time" and persistent activity (Vesuvius type), (4) volcano with lava-lake activity (Kilauea type), (5) volcano with accelerating cycles of eruptions (Asama type), and (6) volcano with both "excitation" and "loading time."

The models consist of mutually independent Poisson processes with rate parameters independent of time. Explicit expressions for the density function of a repose and its rate function, as well as other characteristic functions, therefore can be derived. The concepts of serial density function and serial eruption rate function are introduced for model 5 because it has a distinctive (culminating) eruption in each cycle. These two functions depend on the serial number of a repose, counted from the previous culminating eruption.

It is concluded that rather simple Markov models can describe the volcanic patterns as observed.

## INTRODUCTION

Although every volcano has an individual repose-period pattern, there are, nevertheless, a number of general types of patterns (Wickman, 1966a, 1966e). The observed patterns result from the interaction of many factors; some deterministic and some stochastic. The relative importance of the individual factors cannot be evaluated at present; thus, it is useful to examine simple extreme models and see how well they describe the essential features of the observed patterns. This paper is concerned with the use of simple Markov models, and in this sense it is a continuation of a previous paper (Wickman, 1966e).

## GENERAL FEATURES OF THE MODELS

The models are of the renewal type with time-independent parameters, a reasonable choice in a situation where the functional time dependence of the parameters is unknown. All models are supposed to approximate the pseudostationary patterns that can be determined for volcanoes with a long life. Some of these volcanoes show sudden breaks in their activity patterns. Such situations will be treated as consisting of a number of subpatterns, and changes in their overall behavior will be described as transitions between subpatterns.

The birth process (new volcanoes) will not be treated, because the total recorded number of examples is small. The death process (extinction) of volcanoes is excluded also, because, at present, there is no method to distinguish a dormant volcano from a recently extinct one.

The following terms, notations, and abbreviations will be used: The states of volcanic activity are the eruption state $E_e$ and the persistent activity state $E_p$; for brevity, the latter will be called the p.a. state. Note that this is a classification of states and therefore consistent with the established volcanologic classification, which lists the persistent activity as belonging to the repose period. The lava-lake state will be regarded as a variety of the p.a. state. The inactive portions of a repose period correspond to one or several repose states $R_0$, $R_1$, $R_2$, ... . All rate parameters of transitions between states will be called $\lambda$ with an attached subindex, which has no special reference to the state involved.

Whenever possible, in order to simplify the notation, the same letter symbol will be used both for the probability of a state and for the state itself; for example, $E_e$ will indicate both the eruption state and the probability for its occurrence. The same symbols, but with lowercase letters, will be used for the Laplace transform of the probabilities; for example $r_1$ is the transform of the probability $R_1$. Special notations will be used only to avoid confusion. The mathematical derivations of the functions studied are available from the author.

Volcanologic data usually consist of registrations of the duration of eruptions and reposes. Expressions for the probability of such events are derived for each model. A useful concept is the age-specific eruption rate $\phi(t)$, or the eruption rate. This function is defined in Wickman (1966a, p. 298); it is the ratio of the density function and the survivor function corresponding to a given probability distribution. Approximately, it is the probability of an almost immediate eruption when the repose is known to have the duration (age) t. (The concept of the serial eruption rate is given later.)

The rate parameters are given in the unit month$^{-1}$; therefore it is not necessary to state it explicitly in the examples. The mean will be indicated by $\mu$, the variance by $\sigma^2$, and the standard deviation by $\sigma$. For each model a single set of numerical parameters is used for all numerical examples, including the Monte Carlo simulation.

Parameters may be characterized as "large" or "small," referring to the month as the time unit. It is also necessary to characterize the parameters relative to each other. The following notation will be used: $\lambda_i \approx \lambda_k$ both are about equal; $\lambda_i >> \lambda_k$, $\lambda_i$ is larger than $\lambda_k$; and the new symbol $\lambda_i \cdot > \lambda_k$, which indicates that $\lambda_i$ is substantially larger than $\lambda_k$. Generally, the symbol will be used if a parameter is about 2 to 100 times larger than the other one.

The models should be regarded only as examples of possible models; the patterns of real volcanoes may be described equally well or better by other models. An effort has been made to make the models simple, but they may not represent the simplest model giving a particular pattern. The models have been constructed with a specific physical situation in mind, but it must be realized that the parameters of the models allow many physical interpretations.

## SIMPLE TWO-STAGE VOLCANO WITHOUT PERSISTENT ACTIVITY (FUJI TYPE)

A diagram of the model is shown in Figure 11.1; the essential features are $\lambda_2 \cdot > \lambda_3$ and $\lambda_4$ small. The eruptions are assumed to be short, thus, $\lambda_1$ is large. It is similar to a model previously studied (Wickman, 1966e). A Monte Carlo simulation with $\lambda_1 = 30$, $\lambda_2 = 0.03$, $\lambda_3 = 0.001$, and $\lambda_4 = 0.0003$ is shown in Figure 11.2. Long reposes alternate with periods of frequent eruptions.

The corresponding density function of a repose is shown in Figure 11.3. The shape of the curve is of the exponential type, but the temporal de-

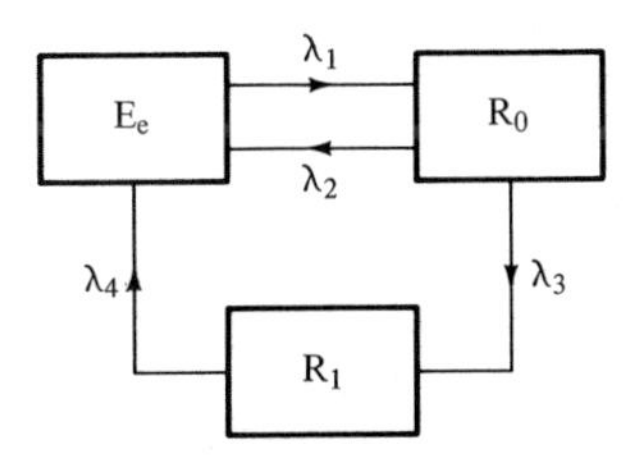


FIGURE 11.1 Diagram of model for simple two-stage volcano without p.a. (Fuji type). $E_e$ = eruption state; $R_0$ and $R_1$ = repose states; $\lambda_1$ to $\lambda_4$ = rate constants for transitions.

velopment is judged better from Figure 11.4, which gives the eruption rate. The probability of an eruption is greatest immediately after an eruption and then decreases to a limiting value.

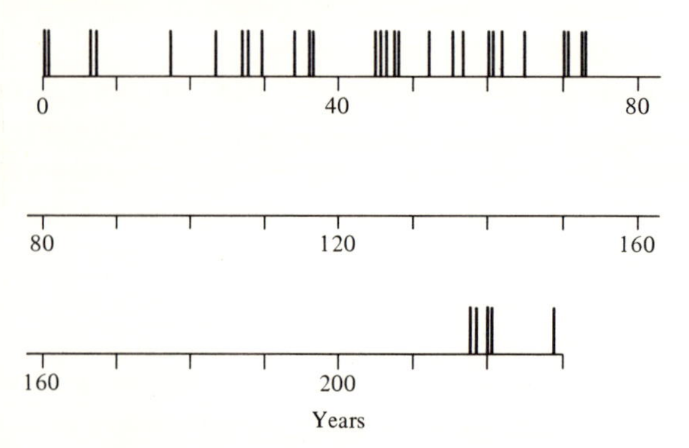


FIGURE 11.2 Monte Carlo simulation of model in Figure 11.1. There is long repose between years 73 and 218; two active periods are shown. For rate constants, see text.

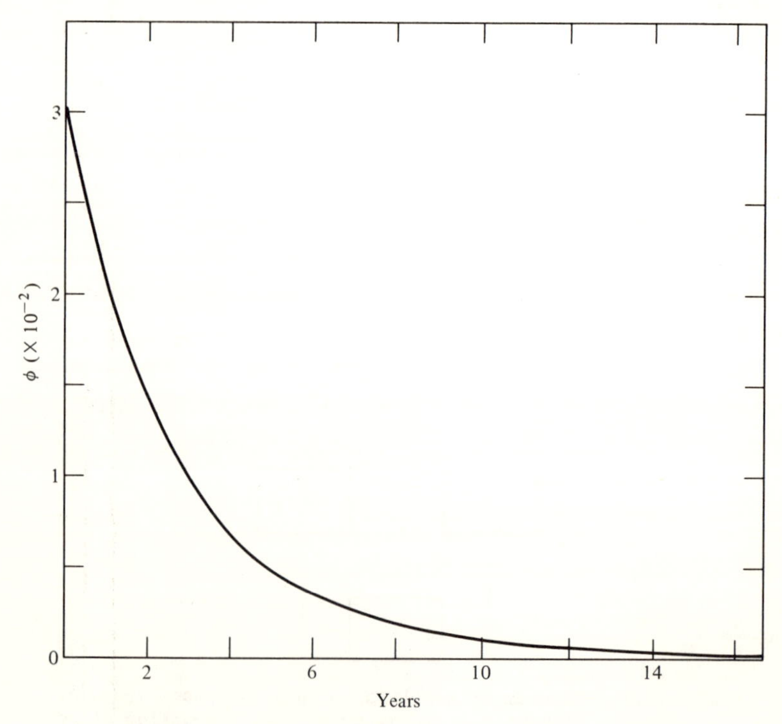


FIGURE 11.3 Density function of duration of reposes; rate constants as in Figure 11.2.

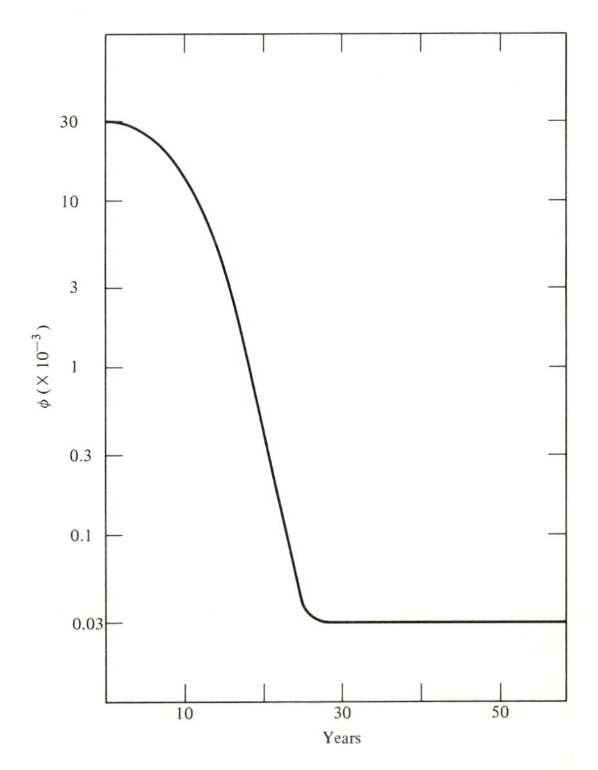


FIGURE 11.4 Eruption rate function $\phi$ with rate constants of Figure 11.2. Abscissa gives age (duration) of repose; ordinate $\phi$ in $10^{-3}$ $months^{-1}$.

The duration of a period of frequent eruptions is the time interval between two consecutive transitions $R_1 \to E_e$ and $R_0 \to R_1$. A long repose consists of the sum of the time consecutively spent in $R_0$ and $R_1$. The numerical parameters give

| | $\mu$ (year) | $\sigma$ (year) |
|---|---|---|
| Period of frequent eruptions | 83 | 83 |
| Long repose | 360 | 275 |

The eruption pattern of this model seems to be rather common among volcanoes, of which the Japanese volcano Fuji (Huzi) is an example. An interesting point about the model is that its eruption statistics will label the volcano as simple Poissonian, if the record is short and only refers to a period of frequent eruptions. Mauna Loa is possibly a natural example, but the present repose must continue for at least another 20 years or so before this conjecture can be taken seriously.

Physical interpretations of the model and its parameters are hypothetical at present. Possibly significant is the observation that tectonic activity in an area of fault lines seemingly moves around, along the lines of weakness in a cyclic pattern. This is referred usually to the building up and release of strain energy. The long reposes could depend on this cyclic pattern of tectonic activity. Another reason, possibly related, could be that frequent eruptions temporarily deplete the magma of a separated gas phase and, at the same time, the conduit may weld together. The gas pressure then can build up again, but it will take time.

However, eruption patterns are only symptoms of the processes going on in the deeper portions of the volcanic structure. Symptoms rarely are unique, as many different processes may give the same pattern.

## VOLCANO WITH "LOADING TIME" AND DORMANCIES (HEKLA TYPE)

The activity of the Icelandic volcano Hekla has inspired this model. Long periods of dormancy alternate with periods of activity, which may persist for perhaps a millennium or more. The reposes during an active period are usually long, as much as a century or longer; short reposes are rare. An active period and the dormancy following it will be termed a "cycle."

A diagram of the model showing two cycles is given in Figure 11.5. The duration of an eruption of Hekla is on the order of months to about a year. The parameter $\lambda$ is thus typically somewhat smaller than 1; all other parameters are necessarily small. It is essential that $\lambda_2 \cdot > \lambda_4$ but the magnitudes of the other parameters are not critical; the pairs $\lambda_2, \lambda_3$ and $\lambda_4, \lambda_5$ are assumed to be of about the same magnitude.

A Monte Carlo simulation is shown in Figure 11.6, where, for simplicity, the same parameters have been used for the two active periods. The figure shows the end of one active period, one dormancy and the active period following it. The parameters are $\lambda_1 = 0.2$, $\lambda_2 = 0.004$, $\lambda_3 = 0.003$, $\lambda_4 = 0.0003$, and $\lambda_5 = 0.0002$. The values have not been selected with the purpose of obtaining the best fit to the observed activity of Hekla. Rather, the figure only illustrates the general behavior of the model. For example, the existence of two reposes shorter than 20 years during 500 years of activity is a feature not observed for Hekla, and depends to some extent on the selected parameter values.

A simple variation of the model would be to permit transitions with a small rate parameter. The present model shows the essential features and therefore has been used.

A repose during the active period consists of the time spent in states $R_0$ and $R_1$. The density function of the duration of a repose is plotted in

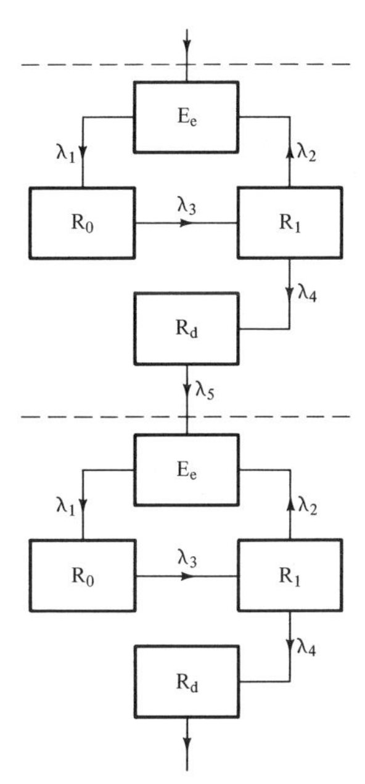


FIGURE 11.5 Two cycles of model for volcano with "loading time" and dormancies (Hekla type). Boundary between cycles indicated by dotted lines. $E_e$ = eruption state; $R_0$ and $R_1$ = ordinary repose states; $R_d$ = dormancy state. Rate constants $\lambda_1$ to $\lambda_5$ may differ from cycle to cycle.

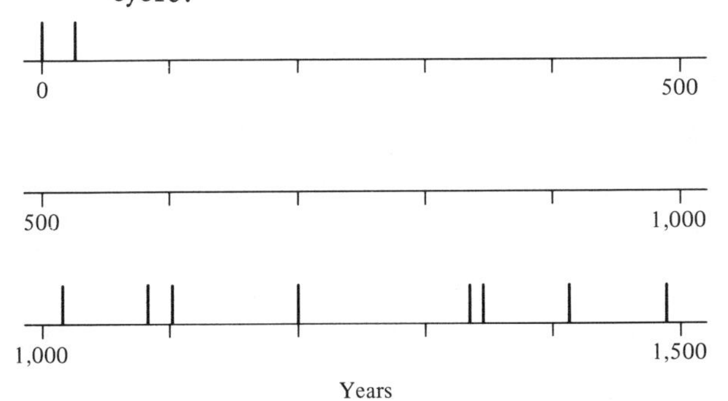


FIGURE 11.6 Monte Carlo simulation of model in Figure 11.5. Dormancy starts after second eruption and lasts for almost one millenium. For rate constants, see text.

Figure 11.7 and is positively skewed. The mode is 24 years. The eruption rate function is shown in Figure 11.8; its general shape is similar to that found for Hekla (Wickman, 1966d).

An active period starts by transition $R_d \rightarrow E_e$ and ends by transition $R_1 \rightarrow R_d$. This transition cannot be observed, so the observational active period ends at the end of the final eruption of the active period, that is, the transition $E_e \rightarrow R_0$.

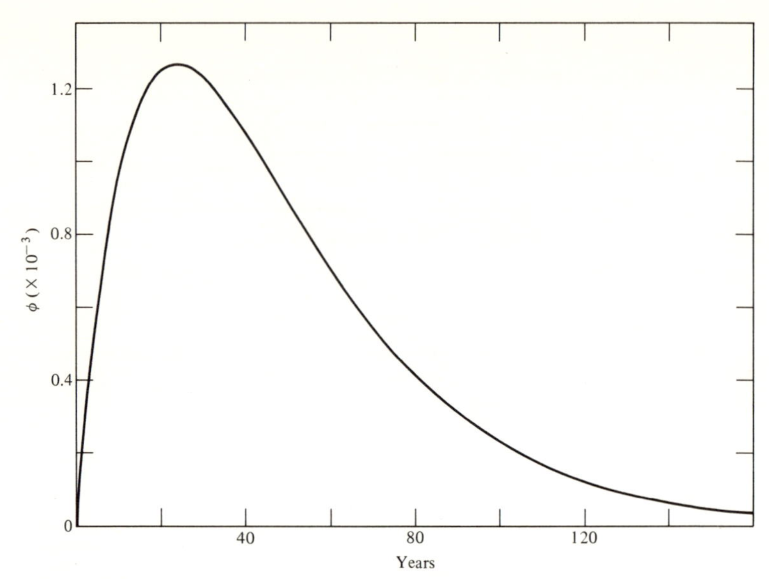


FIGURE 11.7 Density function of duration of repose during active period; rate constants as in Figure 11.6.

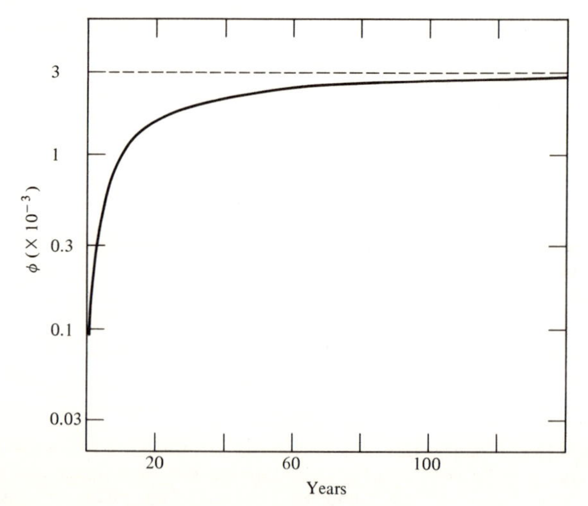


FIGURE 11.8 Eruption rate function $\phi$ with rate constants of Figure 11.6; $\phi$ in $10^{-3}$ month$^{-1}$.

As observed, the dormancy consists of the time spent consecutively in the states $R_0$, $R_1$, and $R_d$. A complete cycle consists of one observational active period and the following observational dormancy. The mean is the sum of their mean values and the variance is the sum of the variances of the two portions. The numerical parameters give

| | $\mu$ (years) | $\sigma$ (years) |
|---|---|---|
| Observational active period | 682 | 669 |
| Observational dormancy | 465 | 417 |
| Complete cycle | 1,147 | 785 |

It has been mentioned that the duration of an eruption of Hekla may be as long as a year. It seems natural, therefore, to assume that the volcano has lost its surplus of gas at the end of an eruption. It takes time to restore the minimum pressure necessary for an eruption, and no eruption can occur during this interval of indefinite duration. The present choice of a constant parameter is convenient but an age-dependent parameter seems more realistic. The small eruption statistics give no real hint. The geologic reasons for the existence of dormancies is supposed to be the same as in the previous model. The main difference is that the corresponding rate parameters are smaller in the present model.

## VOLCANO WITH "LOADING TIME" AND PERSISTENT ACTIVITY (VESUVIUS TYPE)

The normal pattern of the Italian volcano Vesuvius during the last 300 years can be taken as a type example. After a major eruption, Vesuvius shows no activity for a period of time, usually less than a decade. This interval is followed by a period of p.a. that continues indefinitely. Small lava flows may occur during this period. A major eruption starts by the formation of a fissure in the cone, causing the magma level to drop. The duration of an eruption is usually short, ordinarily a few days. It is possible that this pattern changed into another following the 1944 eruption.

The diagram of Figure 11.9 shows the model; it consists of four states $E_e$, $E_p$, $R_0$, and $R_1$, where $E_p$ is the state of p.a. In this model, $\lambda_1$ is necessarily large and $\lambda_2$ is small. Interruptions in pa.a. are short and

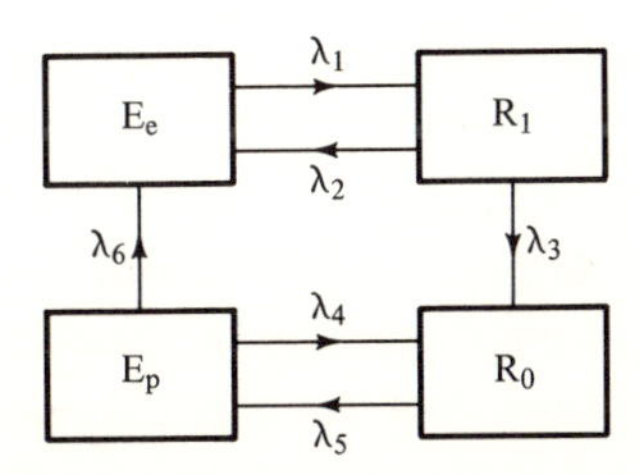


FIGURE 11.9 Diagram of model for volcano with "loading time" and persistent activity (Vesuvius type). $E_e$ = eruption state; $E_p$ = persistent activity state; $R_0$ and $R_1$ = repose states.

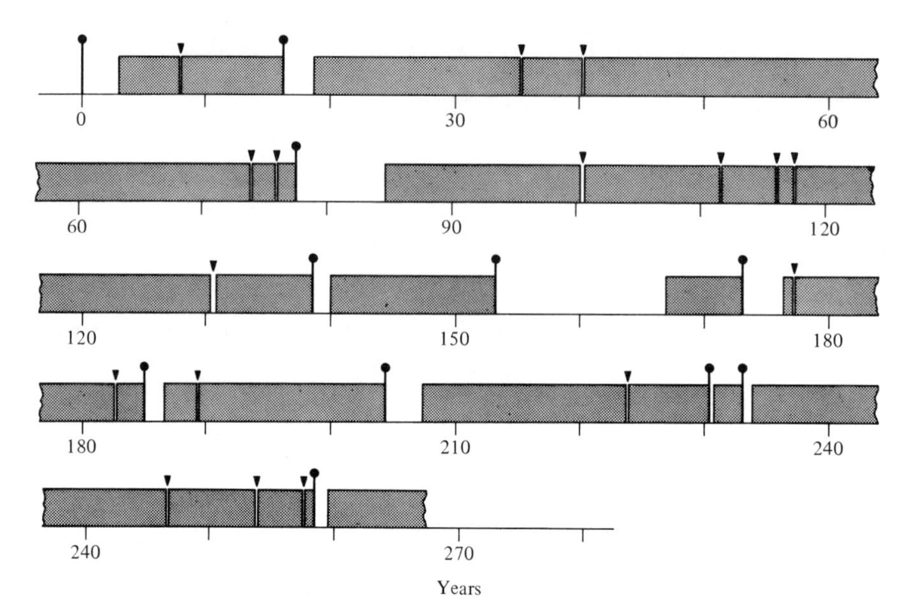


FIGURE 11.10 Monte Carlo simulation of model in Figure 11.9. Periods of p.a. marked in black; major eruptions by dot on line. Inserted triangles mark inactive intervals during periods of p.a. For rate constants, see text.

infrequent, i.e., $\lambda_5$ is large and $\lambda_4$ small. The duration of the inactive interval after a major eruption is determined mainly by $\lambda_3$ in the present model; thus, this parameter must be small, likewise parameter $\lambda_6$; but, one has $\lambda_6 \cdot > \lambda_2$. A more realistic model results if state $R_0$ is halved: one part for the passage from $E_e$ to $E_p$ and the other for inactive intervals during p.a. The eruption statistics do not motivate this elaboration.

A Monte Carlo simulation is shown in Figure 11.10, where the following parameter values have been used: $\lambda_1 = 10$, $\lambda_2 = 0.0001$, $\lambda_3 = 0.03$, $\lambda_4 = 0.03$, $\lambda_5 = 1$, and $\lambda_6 = 0.003$. All eruptions in this simulation are preceded by p.a., indicated in black in the figure.

A repose starts by transition $E_e \rightarrow R_1$, whereas an eruption starts either by transition $E_p \rightarrow E_e$ or by transition $R_1 \rightarrow E_e$. A graph of the density function of duration of a repose is given in Figure 11.11. The function, characterized by a positive skewness, has a maximum of 7.2 years. The corresponding eruption rate is shown in Figure 11.12. Its general shape is similar to that found for Vesuvius (Wickman, 1966d, p. 349).

The mean and variance of the interval between the end of an eruption and the beginning of p.a. are

| | $\mu$ | $\sigma$ |
|---|---|---|
| Repose | 32 years | 29 years |
| Interval between eruption and p.a. | 34 months | 33 months |

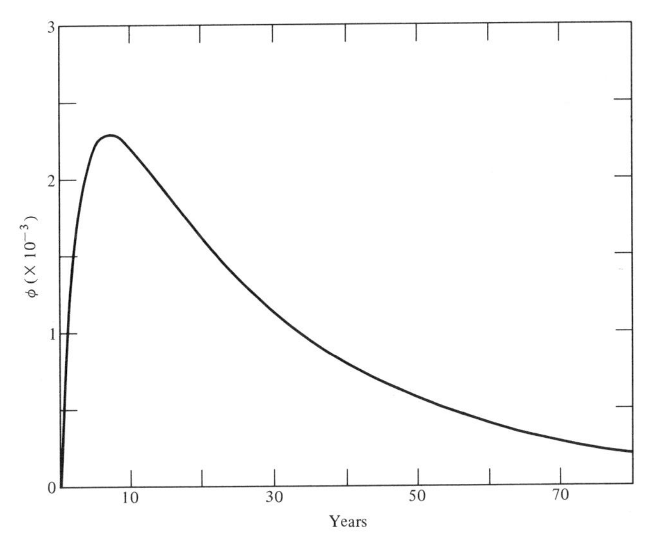


FIGURE 11.11 Density function of duration of repose; rate constants as in Figure 11.10.

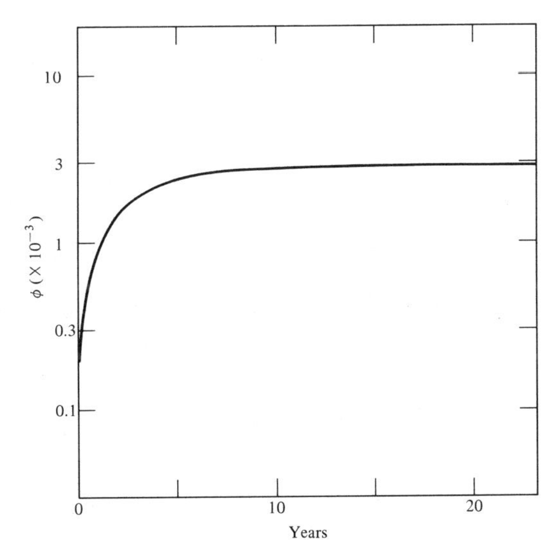


FIGURE 11.12 Eruption rate function $\phi$ with rate constants of Figure 11.10; $\phi$ in $10^{-3}$ $month^{-1}$.

Many variations of this model are possible, for example, one can put $\lambda_2 = 0$, or introduce a state between $R_0$ and $R_1$. Such variations do not change its essential features.

A physical interpretation of the model is the following. An eruption degasses the volcano to such an extent that the likelihood of a new immediate eruption, triggered by some tectonic event, is small (parameter $\lambda_2$). The normal development is, instead, that the gas content increases and the magma level in the conduit rises. The time needed to fill up the conduit to the vent is regarded as being of random length. When the vent is reached by the magma, p.a. starts. Short inactive periods may occur, depending on random disturbances (parameter $\lambda_5$). The p.a. ends by a major eruption, triggered either by the opening of a fissure in the cone or, perhaps, by some deep-seated tectonic event. Both possibilities are regarded as Poissonian, so the parameter $\lambda_6$ is the sum of two parameters, one for each type of event.

## VOLCANO EITH LAVA-LAKE ACTIVITY (KILAUEA TYPE)

The outstanding example is Kilauea. Its activity during the last 150 years can be described as periods of lava-lake activity alternating with periods of "ordinary" volcanic activity.

When existing, the lava lake occupies a portion of the caldera floor and is bounded by a rim. It is difficult to give a satisfactory definition of an eruption during these periods, but intervals when lava is flowing over the lip of the rim may be defined as eruptions. The magma may rapidly disappear into the conduit, draining the lake for a short but indefinite time. A flank eruption starts when a fissure opens on the slope of the volcano, and may be followed by a short inactive interval. The lava-lake activity may persist for many years, or even decades.

Ordinary periods show a different pattern. The eruptions are usually short; flank eruptions are common and p.a. uncommon. A lava-lake period and the following ordinary period will be termed a "cycle."

Figure 11.13 presents a model with many properties characteristic of the real volcano. State $E_e$ is the eruption state; it can be reached either by transition $P_1 \to E_e$ from the preceding ordinary period or by transition $E_p \to E_e$ from the p.a. state. The character of the eruption is determined by the transition that follows: $E_e \to R_1$ will indicate a flank eruption; $E_e \to E_p$, a lava flow over the lip.

Formally, the type of eruption is determined a posteriori, but this is only a convenience, because the states are mutually independent. The description of the eruption state is oversimplified for the following reasons: (1) both types of eruptions are short; and (2) the essential point is to include both types of eruptions in the model, not to give a detailed description of the eruptions. The transition sequence $E_p \to R_0 \to E_p$ indicates the temporary drainage of the lava lake.

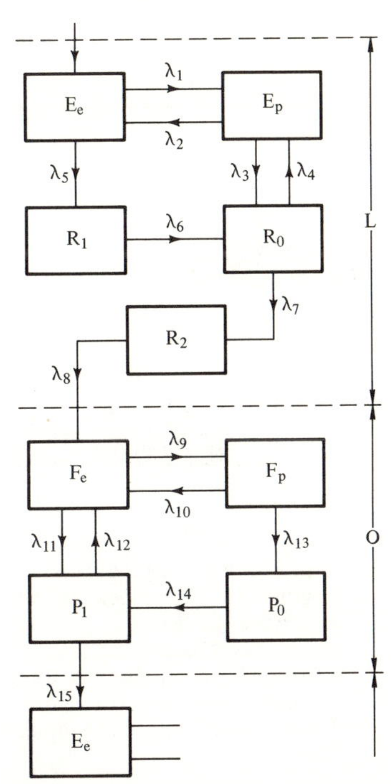


FIGURE 11.13 Diagram of model for volcano with periods of lava-lake activity (Kilauea type). Letters E and F indicate eruption states; P and R, repose states. For details, see text. Lava-lake period is indicated by letter L and "ordinary" period by O. Rate constants for each period may differ from other periods.

The eruption and repose states during an ordinary period have been termed F and P, respectively. No distinction is made between summit and flank eruptions during these periods in order to simplify the model.

We assume $\lambda_1$ large because the eruptions during the lava-lake period are short and $\lambda_1 \cdot > \lambda_5$ because flank eruptions will be made rare. It is also characteristic that $\lambda_2 \cdot > \lambda_3$ and $\lambda_1 \cdot > \lambda_2$; the drainage of the lava lake is short and thus $\lambda_4$ large. The size relation of $\lambda_5$ and $\lambda_6$ is not critical; they may be chosen as $\lambda_5 \approx \lambda_6$. The parameters $\lambda_4$ and $\lambda_7$ are governed by the condition $\lambda_4 \cdot > \lambda_7$. The relative size of $\lambda_7$ and $\lambda_8$ is not known from observations but does not seem critical. The condition $\lambda_{11} \cdot > \lambda_9$ seems valid for the ordinary period. The size relation of $\lambda_{10}$ and $\lambda_{13}$ is chosen as $\lambda_{13} \cdot > \lambda_{10}$; $\lambda_{14}$ does not seem critical, but $\lambda_{12} \cdot > \lambda_{15}$, $\lambda_{11} >> \lambda_{15}$, and $\lambda_{11} \cdot > \lambda_{12}$.

A Monte Carlo simulation is shown in Figure 11.14. For simplicity, the cycles have the same set of parameters: $\lambda_1 = 2$, $\lambda_2 = 0.2$, $\lambda_3 = 0.02$, $\lambda_4 = 1$, $\lambda_5 = 0.3$, $\lambda_6 = 0.3$, $\lambda_7 = 0.3$, $\lambda_8 = 0.03$, $\lambda_9 = 0.1$, $\lambda_{10} = 0.05$, $\lambda_{11} = 2$, $\lambda_{12} = 0.03$, $\lambda_{13} = 0.2$, $\lambda_{14} = 0.1$, and $\lambda_{15} = 0.001$. The letter F

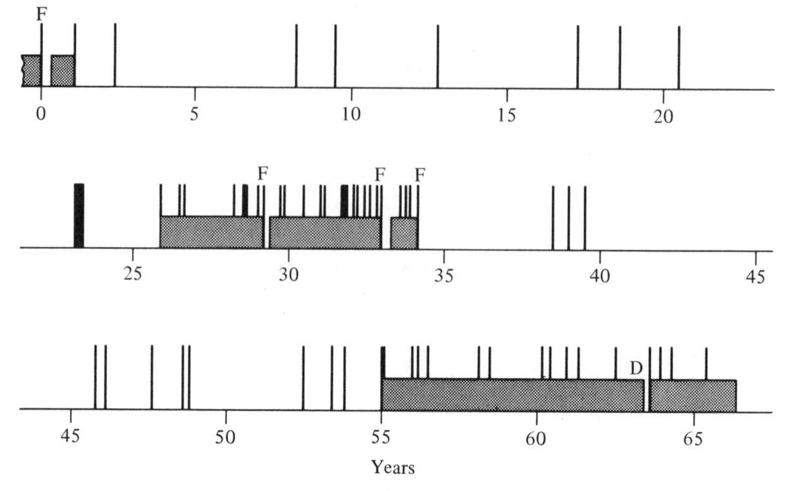


FIGURE 11.14 Monte Carlo simulation of model in Figure 11.13. "Low" black areas indicate lava-lake periods; "high" ones, "ordinary" activity. F = flank eruption during lava-lake period; D = temporary drainage of lava lake. For rate constants, see text.

in the figure indicates a flank eruption and the letter D an interval when the lava lake is drained; both refer only to lava-lake periods. The frequency of flank eruptions during the period around 30 years is uncommonly high.

(Note that these numerical parameters were not selected to describe Kilauea, except in a general way; they were merely chosen to give an idea of the behavior of the model.)

The density function of the duration of a repose during a lava-lake period is shown in Figure 11.15; its general shape is exponential, but the eruption rate curve shows a more complex structure (Figure 11.16). The eruption rate function drops rapidly during the first month or so of the repose, followed by a slower decrease to the asymptotic value of $30 \times 10^{-3}$ month$^{-1}$. If the shape is compared with that of the eruption rate curves shown in Wickman (1966a, 1966d) it must be observed that the ordinate is linear in Figure 11.16 and logarithmic in the papers mentioned. The mean of the duration of a lava-lake period is determined to be about 69 months, or almost 6 years.

The density function of the ordinary period has an exponential shape (Figure 11.17). The eruption rate curve is shown in Figure 11.18; it is almost constant. The mean duration of an ordinary period is found to be 1,033 months, or almost exactly 86 years. The ordinary periods shown in Figure 11.14, therefore, are on the short side.

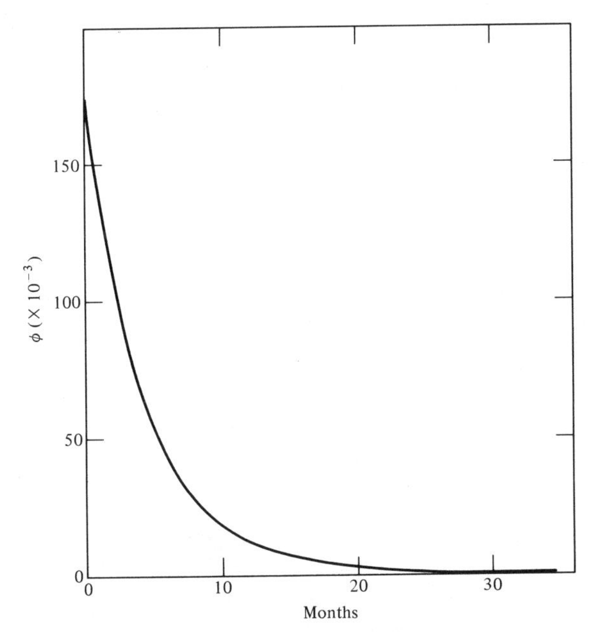


FIGURE 11.15 Density function of duration of repose during lava-lake period; rate constants as in Figure 11.14.

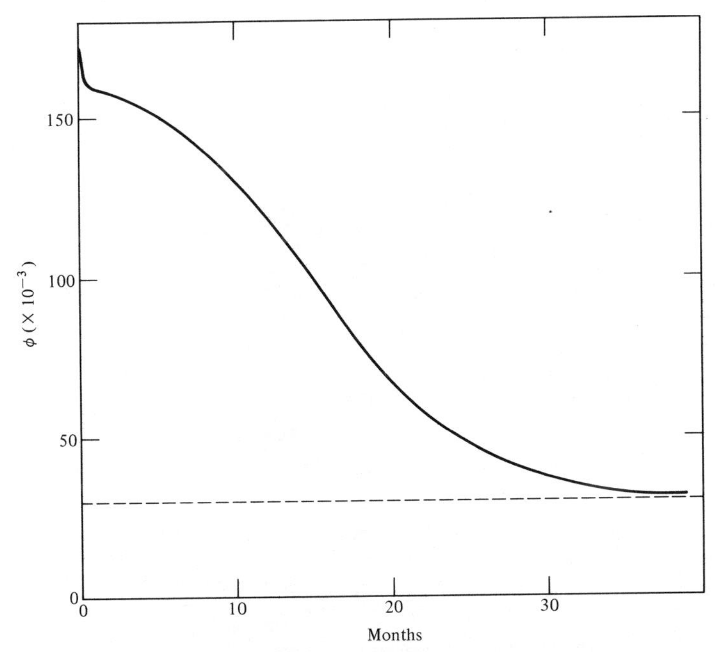


FIGURE 11.16 Eruption rate function $\phi$ for reposes during lava-lake period using rate constants of Figure 11.14; $\phi$ in $10^{-3}$ month$^{-1}$.

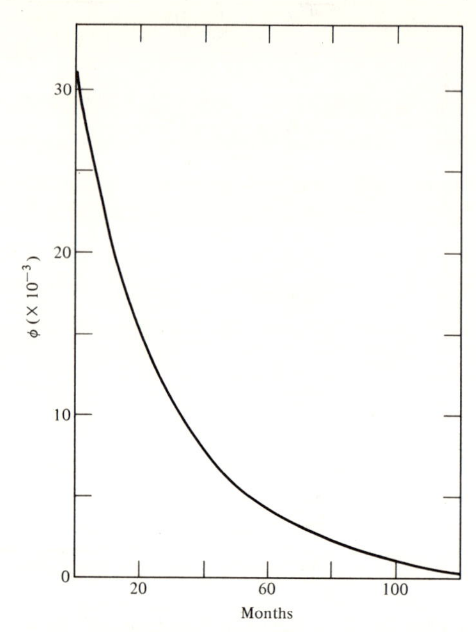


FIGURE 11.17 Density function of duration of repose during "ordinary" period; rate constants as in Figure 11.14.

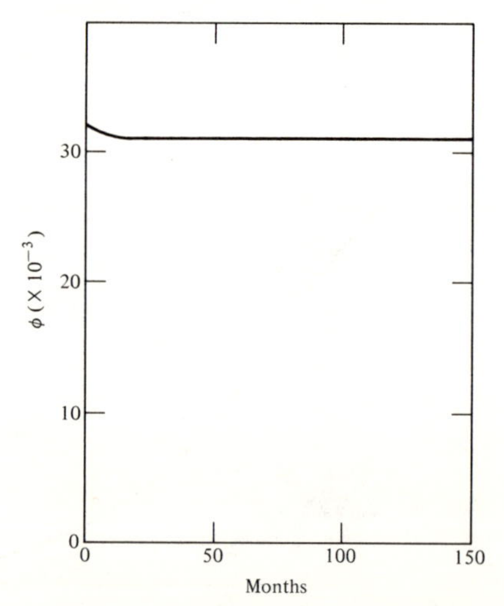


FIGURE 11.18 Eruption rate function $\phi$ for reposes during "ordinary" period using rate constants of Figure 11.14; $\phi$ in $10^{-3}$ $month^{-1}$.

The central issue in the physical interpretation of the present model is to establish those conditions essential to produce a lava lake in a volcano. A low-viscosity magma and a suitable depression are rather trivial aspects. More interesting, perhaps, is the indication that flank eruptions are commoner during ordinary periods than during lava-lake periods. One can hypothesize that the upper magma chamber, which is known to occur in the porous volcanic cone, consists of a set of dikes filled with magma and variable in number and sizes. If the magma is cooled, it will weld the rocks.

Structure of the volcanic cone is mechanically weak and may crack when mechanical forces are acting on it. Flank eruptions may occur if the cracks are close to the surface. In this manner the ordinary period is realized. However, the magma in these surficial portions will cool more rapidly and weld the structure. The repair will work its way upward and when, after perhaps a number of setbacks, the structure is repaired, the condition for the formation of a lava-lake exists. The lava lake will exist until new cracks appear, probably occurring along the fissure zone through the volcano. It is possible that this chain of events can be regarded as Markov to the first approximation.

The various states during the lava-lake period can be regarded as the result of the superposition of a number of factors, such as Jaggar's old idea with tides, smaller shocks, opening of small dikes, etc. They can be described, then, as in the model.

## VOLCANO WITH ACCELERATING CYCLES OF ERUPTIONS (ASAMA TYPE)

The phrase is not particularly descriptive and Asama is not a good example, but the pattern is difficult to summarize in a few words. The typical pattern would consist of separate cycles of undetermined length and the boundary between two cycles distinguished by a special type of eruption, for example, a pyroclastic flow. The first repose after such a culminating eruption is usually longer than the average; short reposes then follow with increasing frequency, ending in a new culminating eruption.

It is difficult to judge whether this idealized type exists in nature, partly because the historic record of most volcanoes is too short. However, one portion of the recorded history of the Japanese volcano Asama gave the idea for the model. According to Aramaki (1963), Asama seems to have produced pyroclastic flows on two occasions (1281 and 1783). The period between 1281 and 1783 follows the pattern sketched; the first repose was 250 years. The pattern after the 1783 eruption seems to have a different course. The model is a simplification because Asama also has shown a special type of p.a. for long periods of time.

Many models can generate the desired pattern. One method is to use time-dependent rate parameters; another, which will be used here, is to use a series of repose states characterized by increasingly larger but time-independent rate parameters. A diagram of the model is shown in Figure 11.19.

One may ask whether this model has been chosen solely for convenience and in order to conform with the previous models treated in this paper, or whether some more fundamental reason exists. The answer depends to a great extent on one's view about the temporal development of the physical conditions within a volcanic structure. If the conditions mainly change stepwise, the present Markov model is more realistic than a model based on parameters that increase continuously during a cycle. Although it is true that eruptions are abrupt changes in the physical conditions, seen in the proper scale, they may be insignificant. If this is true, the continuous model is more proper. A mixed model is probably the most realistic one.

Whatever the answer, the model of Figure 11.19 has interesting properties. The number of states has been kept small in order to limit the number of parameters, and for the same reason the eruption states $E_1$, $E_2$, and $E_3$ are assumed to be short-lived so that they can be regarded as instantaneous if compared to the duration of a repose.

The state $E_0$ refers to the distinctive eruption; it is longer than the other eruptions, but for convenience $\lambda_0$ is large. The first repose has to

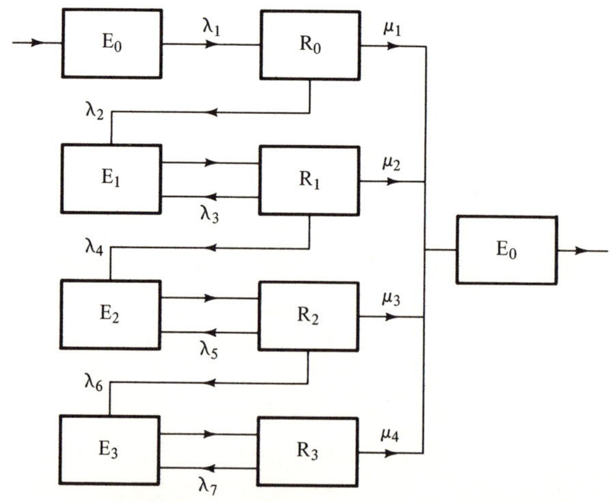


FIGURE 11.19 Diagram of model for volcano with accelerating cycles (Asama type). $E_0$ = culminating eruption state; other E's = ordinary eruption states; R = repose states; $\lambda$ and $\mu$ = rate constants. For details, see text.

be long and therefore $\lambda_1$ is small. The liklihood of an immediate repetition of a culminating eruption is assumed to be small, that is, $\mu_0$ is small and $\lambda_1 \cdot > \mu_0$. Three mutually independent events are possible in $R_1$: their respective parameters are $\lambda_2$, $\lambda_3$, and $\mu_1$. The relative magnitudes of $\lambda_2$ and $\lambda_3$ cannot be judged but are assumed not to be too critical; in the example, $\lambda_2 \approx \lambda_3$. However, it seems natural that both $\lambda_2$ and $\lambda_3$ are substantially larger than $\mu_1$ and $\mu_1 \cdot > \mu_0$. The same can be said for $\lambda_4$, $\lambda_5$, and $\mu_2$, as well as for $\lambda_6$ and $\mu_3$. The overall picture is one of increasing parameters.

A Monte Carlo simulation of a few cycles is shown in Figure 11.20, using the following parameters: $\lambda_0 = 2$, $\lambda_1 = 0.0005$, $\lambda_2 = 0.0013$, $\lambda_3 = 0.001$, $\lambda_4 = 0.004$, $\lambda_5 = 0.003$, $\lambda_6 = 0.01$, $\mu_0 = 5 \times 10^{-5}$; $\mu_1 = 30 \times 10^{-5}$, $\mu_2 = 70 \times 10^{-5}$, and $\mu_3 = 300 \times 10^{-5}$. The culminating eruptions are marked by a dot; black areas indicate intervals when eruptions occur frequently, that is, when most reposes are shorter than one year. It is only the first long repose that shows clearly in the simulation. There are several reasons for this: the particular choice of the numerical parameters, the small number of states, and, perhaps most important, the fluctuations that result if only a small number of eruptions are likely to occur within each cycle.

The eruption rate of this model has a type of memory effect. The first two reposes must be spent in states $R_0$ and $R_1$, respectively. The states of the following reposes are determined by chance, so the third repose can be spent in either $R_1$ or $R_2$ and all the following in one of the

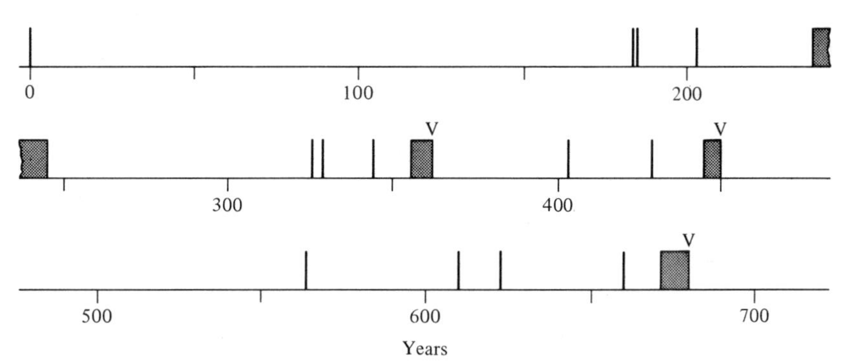


FIGURE 11.20 Monte Carlo simulation of model in Figure 11.19. V-shaped symbol indicates culminating eruption, followed by long repose. Dark areas indicate periods of frequent eruptions (reposes around or less than one year). For rate constants, see text.

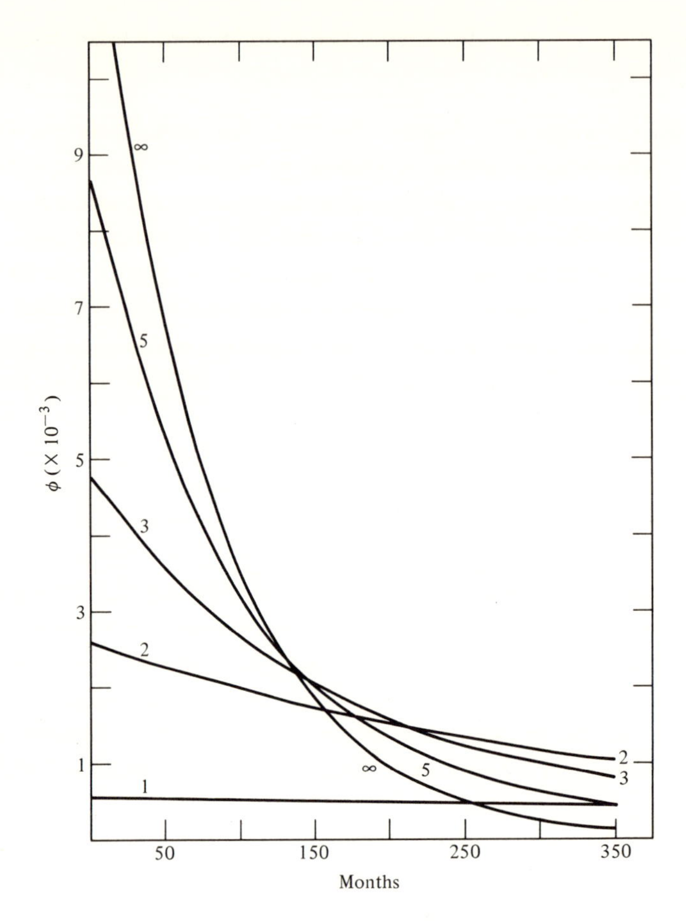


FIGURE 11.21 Serial density function of reposes 1, 2, 3, 5, and ∞ in cycle; rate constants as in Figure 11.20.

states $R_1$, $R_2$, or $R_3$. The special character of the $E_0$ eruption makes it possible to number the reposes of a cycle.

We define therefore a special eruption rate function, which is not only a function of the age of a repose but also of its serial number n in the cycle. This function will be called the serial eruption rate (function) and written as $\Psi_n(t)$. This function is particularly suited for this model and, furthermore, is easy to derive. The corresponding density functions of the different reposes in a cycle are shown in Figure 11.21. The scale of the figure has been selected to show the densities for reposes with $n \geqslant 2$; the density function of the first repose especially does not show much variation. The limiting density function for $n \to \infty$ is shown as well as a few other densities. Some serial eruption rate functions are shown in Figure 11.22. Three are constant: $\Psi_1$, $\Psi_2$, and $\Psi_\infty$. The other ones

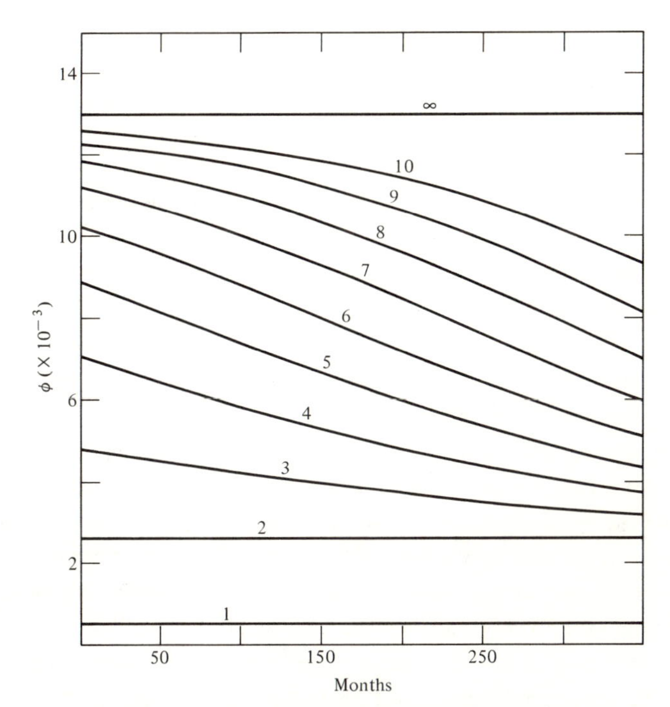


FIGURE 11.22 Serial eruption rate functions $\Psi_n$ for n 1, 2, 3, 4, 5, 6, 7, 8, 9, 10, and ∞. Rate constants as in Figure 11.20; $\phi$ in $10^{-3}$ $month^{-1}$.

approach $\Psi_2$ asymptotically. In this figure the properties of the model are well demonstrated. The mean and variance of a cycle give $\mu = 241$ years and $\sigma = 166$ years.

One physical interpretation of this model is that the culminating eruption empties the volcano of available highly viscous pyroclastic magma. The long repose, then, is necessary to rebuild the source. The magma will have to work its way up through the conduit, that is, some material has to be extruded before it reaches the vent again. The closer the pyroclastic magma comes to the surface, the easier the stepwise upward movement will occur.

## VOLCANO WITH BOTH "EXCITATION" AND "LOADING TIME"

The observed eruption rate function of many volcanoes (Wickman, 1966b, 1966d) seemingly has a minimum at "middle-aged" reposes. This is partly an illusion based on at least two factors. One cause is that many early reports were made by inexperienced casual visitors who may easily have mistaken p.a. for an eruption. If no detailed descriptions of the accompanying phenomena are given, such mistakes cannot be corrected. This factor results in an increase in the number of short reposes. Also, the number of long reposes is necessarily small, depending on the short historic records; consequently, random fluctuations in the observed durations are important. Fortuitously, the few long reposes in a record may have roughly the same duration, sometimes resulting in a higher eruption rate for great repose age.

Even if most or all observed examples can be disqualified for these or other causes, there seems to be no fundamental reason why such eruption rate functions should not exist. It is from this viewpoint that it is interesting to examine a simple Markov model having an eruption rate of this type. The diagram in Figure 11.23 shows the model, consisting of two portions marked by the letters S and L, respectively, in the diagram. A proper choice of parameter values will give mainly short reposes in the S-marked portion and longer ones in the L-marked portion. The volcano will alternate irregularly between the two portions.

It is easy to see how the model can be elaborated, but the present one shows the essential feature and therefore will be used. The duration of an eruption is not particularly interesting in the present situation, so we will make it short, that is, parameters $\lambda_1$ and $\lambda_3$ are large. However, as the model is constructed, these parameters determine the likelihood of a short or a long repose, so they have to be selected correspondingly. The parameter $\lambda_2$ is large by definition, whereas $\lambda_4$ and $\lambda_5$, or at least one of them, must be small.

A Monte Carlo simulation is shown in Figure 11.24 with the following parameters: $\lambda_1 = 2$, $\lambda_2 = 0.3$, $\lambda_3 = 2$, $\lambda_4 = 0.02$, and $\lambda_5 = 0.02$. The spotty appearance of short reposes is evident.

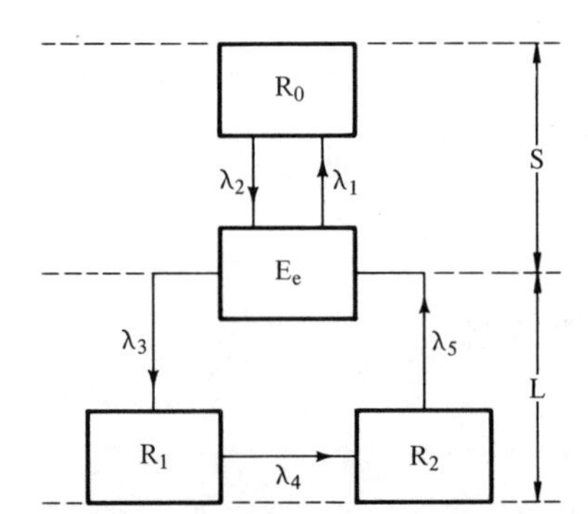


FIGURE 11.23 Diagram of model for volcano with both "excitation" and "loading time." $E_e$ = eruption state; $R_0$, $R_1$, and $R_2$ = repose states; S = "short" reposes; L = "long" reposes.

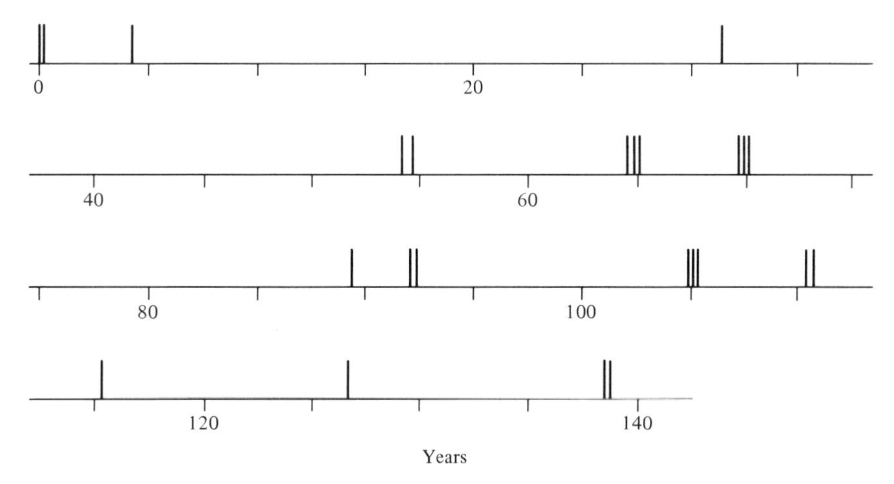


FIGURE 11.24 Monte Carlo simulation of model in Figure 11.23. For rate constants, see text.

The density function of the duration of a repose is shown in Figure 11.25 in two scales. The one in small scale does not show the minimum in the density function clearly, so an enlarged portion has been included. The value of the density function for t = 0 is 0.15, which is outside the figure.

The corresponding eruption rate function is shown in Figure 11.26. The asymptotic value is $20 \times 10^{-3}$ $month^{-1}$; the curve shows a well-developed minimum.

If natural counterparts to this model exist, they probably occur where the physical conditions are related to those of the Fuji type. The main differences are their respective time scales and the mixing of the two types of eruptions. In the Fuji type, the long reposes almost certainly come singly.

A simple physical interpretation of the model is based on the occurrence of two situations. One situation is where eruptions are likely but, at the same time, conditions are not stable enough to produce p.a., the magma level is stabilized at various levels fairly deep in the conduit, or a near-surface magma chamber may develop. The other situation is that the volcano gets "exhausted" by loss of gases or for other reasons. In order to restore the proper conditions for an eruption, the volcano has to pass several stages, similar to a volcano of the Hekla type. The major difference here is that the time scale is shorter.

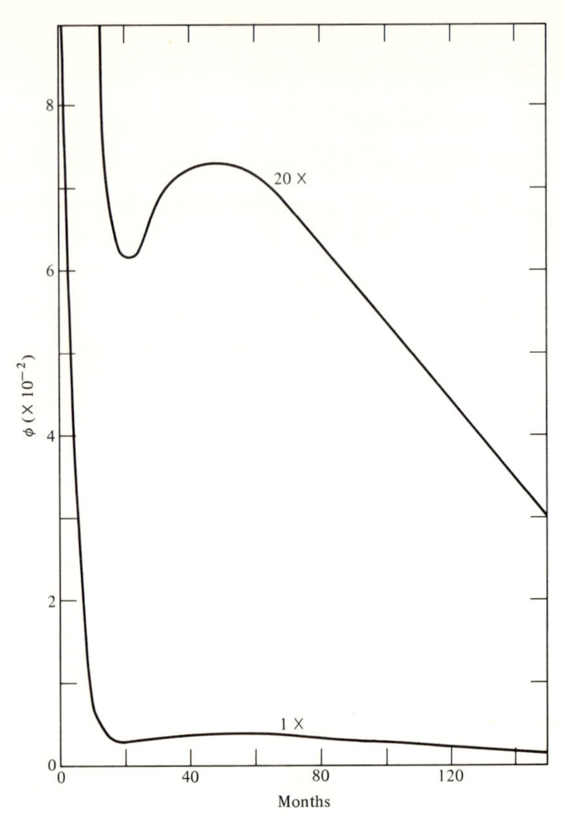


FIGURE 11.25 Density function of duration of repose. Shape is shown at 20 x magnification; rate constants as in Figure 11.24.

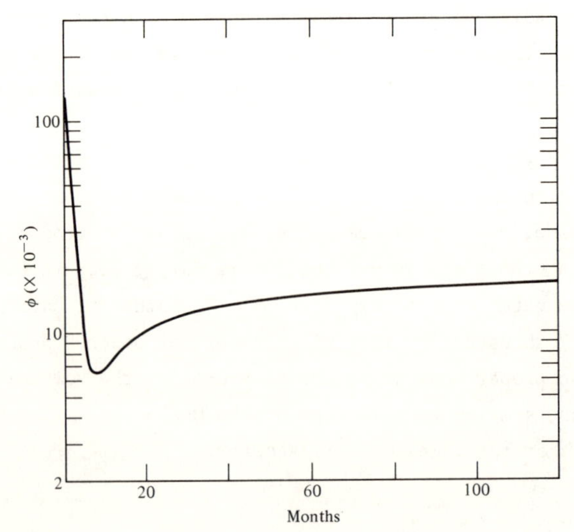


FIGURE 11.26 Eruption rate function $\phi$; rate constants as in Figure 11.24; $\phi$ in $10^{-3}$ $month^{-1}$.

## GENERAL DISCUSSION AND CONCLUSIONS

The models have not been tested against the observed records of volcanoes by some strict statistical procedure. Such tests have purposely been omitted, because the historic records suffer from many weaknesses (Wickman, 1966a, 1966b, 1966c, 1966d). The records are short, and the possibility exists that in some records eruptions are missing or have been added mistakenly. Some models also are idealized only from a portion of the total record. A statistical test would be nothing more than a numerical exercise.

The models have to be accepted or rejected for other reasons, mainly geologic. Volcanic regions are directly related to specific, large-scale tectonic situations, that is, they are typically deterministic phenomena. This fact does not exclude or contradict the possibility that the occurrence of eruptions of an individual volcano can be trested as a random phenomena. Among other things, an eruption of a specific volcano depends on the availability of magma and on a triggering mechanism that functions. Both factors, as well as their mutual interaction, are complex. It is to be expected, therefore, that the occurrence of eruptions can be treated as a stochastic process.

The six models have shown that simple Poisson processes can be combined into simple Markov models that, qualitatively at least, can describe complex repose patterns. Because all models refer to specific patterns, it is natural to ask whether a master model, including the six models as special examples, would be preferable. At present, this is not the situation, because the main interest is not in generalities but in the behavior of individual volcanoes. But it is clearly important to examine to what extent the six models can be interpreted as variations of the same eruption mechanism. We shall examine this question briefly.

Persistent activity is regarded as a repose, and may occur if the magma is not too viscous and the conduit is open. It will occur if the quasiequilibrium surface level of the magma happens to coincide with some level in the accessible crater. The p.a. usually stops at the occurrence of some minor or major event in the volcanic structure. If the event is minor, quasiequilibrium and p.a. soon will be restored, whereas a bigger event may result in more drastic changes in the pattern. In the example of Kilauea, it seems possible that a submarine flank eruption could have caused the present ordinary period of activity. The disturbance may be so strong as to permanently change the pattern.

Regular eruptions occur when there are mechanical failures in the volcanic structure. The cause of the failure may differ but typically it is the result of tectonic movements, increased load, or fatigue of some portion of the volcanic structure. In the models the duration of a repose therefore is regarded mainly as dependent on mechanical failures and the availability of magma.

Magma is either always available or the volcano needs "reloading." If magma is available, no particular state is needed, whereas a loading interval needs at least two states. In the first of these states transitions to an eruption state are forbidden. Examples of this type are shown in Figure 11.5 and Figure 11.23.

An eruption is triggered by some mechanical failure in the volcanic structure, but the construction of the model must take into consideration the type of failure and how it is repaired. For example, if the part that failed is easily repaired to its earlier strength, then in the model, one can use a simple alternation between an eruption state and a repose state. If the repair is not likely to be successful, two repose states are necessary: one for the situation when a new failure is likely and one for the successful ones. It also happens that failures become less likely, often during a repose, for reasons such as welding of the conduit or, with less likelihood, tectonic movements. When this happens, an additional repose state is necessary. Examples of these principles can be found in most of the models.

A major factor in determining the number of repose states is how many different time scales are necessary to describe the pattern. The spectrum is wide, from a dormancy of several millennia to short episodes of p.a.

Simple models similar to those described can give a general description of the different repose patterns. It seems natural, then, to conclude that more realistic and, therefore, elaborate, models would be able to describe the activity more accurately. The major feature of the models, which has to be changed, is the assumption that the volcanologic conditions are constant; in reality every volcano is a transient phenomenon and develops accordingly. But in many situations volcanoes seem to exist for many hectomillennia or even megennia.

At present, however, another more difficult problem seems urgent: to describe the interaction between the size (energy) of an eruption and the following repose or perhaps reposes. In other words, one may fact problems that need non-Markov models from the outset.

## REFERENCES

Aramaki, S., 1963, Geology of Asama volcano: Jour. Fac. Sci. Univ. Tokyo, sec. II, v. 14, p. 229-443.

Wickman, F. E., 1966a, Repose-period patterns of volcanoes, I. Volcanic eruptions regarded as random phenomena: Arkiv for Mineralogi och Geologi, Bd 4, Hafte 4, p. 291-301.

Wickman, F. E., 1966b, Repose-period patterns of volcanoes. II. Eruption histories of some East Indian volcanoes: Arkiv for Mineralogi och Geologi, Bd 4, Hafte 4, p. 303-317.

Wickman, F. E., 1966c, Repose-period patterns of volcanoes. III. Eruption histories of some Japanese volcanoes: Arkiv for Mineralogi och Geologi, Bd 4, Hafte 4, p. 319-335.

Wickman, F. E., 1966d, Repose-period patterns of volcanoes. IV. Eruption histories of some selected volcanoes: Arkiv for Mineralogi och Geologi, Bd 4, Hafte 4, p. 337-350.

Wickman, F. E., 1966e, Repose-period patterns of volcanoes. V. General discussion and a tentative stochastic model: Arkiv for Mineralogi och Geologi, Bd 4, Hafte 5, p. 351-367.

# Index